HAMLYN

Animal Coloration

The nature and purpose of colours in vertebrates

Animal Coloration

The nature and purpose of colours in vertebrates

Ivan Heráň

HAMLYN

London · New York · Sydney · Toronto

Translated by Stuart Amor
Consultant editors Sheila Shaw B.Sc. and Peter Shaw B.Sc.,
Dip. Ed., M.I. Biol.

Published by the Hamlyn Publishing Group Limited
London · New York · Sydney · Toronto
Astronaut House, Feltham, Middlesex, England

ISBN 0 600 30303 9

Printed in Czechoslovakia
3/09/01/51-01

CONTENTS

Introduction 9
The Nature of Animal Coloration 13
Colours for Recognition 21
Patterns and Colours 28
Colour Changes 38
Anomalies in Coloration 46
Coloration and Genetics 60
Sexual Dimorphism in Coloration 69
Age Differences in Coloration 77
Seasonal Variations 85
Protective Coloration 92
Camouflage 96
Warning Colours 111
Disguise and Mimicry 116
Colours as Part of Protective Equipment 124
How Animals See 130
The Effectiveness of Coloration 136
Coloration for Communication 140
Colour and Pattern in the Animal Kingdom
  148
Unanswered Questions 156

INTRODUCTION

At the present time man is subject to more rapid changes in his way of life and has a greater impact on his environment than at any time in his history and this is clearly reflected in his current interest in and concern about his surroundings.

Many people, especially city dwellers, are more or less permanently excluded from direct contact with the natural world — a world which is continually shrinking under the relentless advance of civilization and an almost explosive increase in the world's population.

To some it might seem that the new environment created in modern times is more beneficial to man and that its advantages can satisfy all his needs. However, experiences in many countries show that even today nature has preserved its attraction for man and that all the comfort provided by modern technology cannot suppress in him the desire for immediate contact with living creatures. Most of us are no longer capable of those things which came quite naturally to our ancestors; moving about in nature and reading all the signs with which it speaks to us. This is apparently the source of that great and lasting enthusiasm for popular literature and television programmes on the world of nature.

Nowadays the study of animals is pursued in many different ways and information about the appearance and distribution of living creatures has long since ceased to be the only aspect of our knowledge of them. Popular literature can no longer offer merely a stereotyped handbook but must provide a more detailed view of the living world. An endeavour to learn about and understand the way of life of living creatures, as undertaken by morphologists, ecologists, geneticists or ethologists, is ceasing to be a mere indulgence and is becoming a means of understanding our own existence, biological aptitudes and abilities.

Man lives primarily in a world of colours and forms. His sense of orientation within his environment is based on visual impressions, by means of which he directs the greater part of his activities and upon which is also built the foundation for a substantial proportion of human culture and art. For someone who loses his sight the whole world round him suddenly changes. He becomes helpless. He gropes, trying with his other senses to perceive what till then had been self-evident. A familiar piece of countryside can become an alien environment at night for the senses of sound and smell are incomparably less effective than visual perception. Without his sight man becomes, at least at first, uncertain and almost helpless. This is why he attaches such importance to visual stimuli and also why he pays particular attention to striking colours and patterns.

The world of animals is literally a riot of such colours and forms. It contains almost all known colours, simple coloration, brightly coloured patterns and bizarre forms. There are species of animals where the rich coloration of the male differs from that of the inconspicuous female; species where drab parents have brightly coloured young. There are forms in which the coloration changes several times during a lifetime and there are animals which change colour according to the season of the year or according to the environment in which they are moving.

Human beings marvel at this variety and at the same time are prompted to wonder what produces it all and what is its significance for the animals? Some of the answers are known but there is much that can only be guessed at. The peculiarities and curiosities in the coloration of living creatures are so numerous that not all of them can be reduced to rules and laws. In addition to this, many ideas disintegrate the moment one stops looking at the function of coloration with human eyes and begins to assess it according to the aptitudes and biological characteristics of the animals that bear it or the milieu in which the coloration is to have its effect.

This book is not a specialist monograph that can answer all these questions. It does not and cannot provide a complete picture of the variety and the function of coloration in the animal world. Rather it is a series of samples and examples and it points out the wider context that escapes us in a cursory study.

In any case, the desire for knowledge is for most people not the main factor in arousing their interest in animals; more often it is a hobby which only subsequently leads to a more profound understanding of the various aspects of animal life and appearance. The aim of this book is primarily to enhance this human pleasure in the things that interest us and at the same time help to satisfy our aesthetic requirements and maintain the mental balance which is so necessary for life in this hurly-burly world.

THE NATURE OF ANIMAL COLORATION

In order to explain the production of animal coloration in such a way as to avoid superficiality and a mere restatement of generally known facts, it is necessary to limit somewhat the definition of the term 'animal'. An amazing similarity has been found in the principles underlying coloration as well as in the actual patterns and colours of such widely different heterogeneous groups as vertebrates and invertebrates. The principles can therefore best be considered in the group of animals most closely related to man, both in body structure and the mechanism of visual perception, namely the vertebrates, and thus vertebrates will be the main consideration of this book.

The coloration of vertebrates is for the most part produced by pigments which are contained in special cells called chromatophores. These cells are quite large and require only slight magnification to be visible, for example, in the scales of carp and other fishes. Under the microscope the pigment within each cell has the appearance of tiny stars or tufts of hair, according to the way in which it is dispersed. If the pigment is dispersed evenly, most of the light entering the cell is absorbed and only a comparatively small amount is reflected, making the colour appear dark. When the pigment is concentrated at the centre of the cell, the reverse situation obtains and the scale appears light in colour. The movement of the pigment in the cell, which is influenced among other things by hormone action, makes possible the so-called adaptive colour change

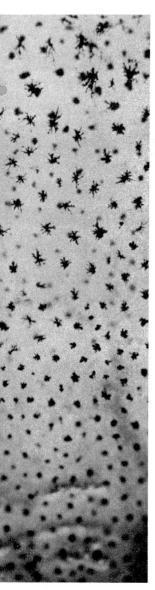

When the scales of the carp *(Cyprinus carpio)* are examined under a microscope, the pigment cells look like tiny stars. (At the top is the edge of the scale and below is the central part of the scale, magnified ten times.)

in animals. The branched pigment cells themselves, however, have a permanent shape and cannot move.

The most common pigments in the pigment cells are melanins. These are dark, generally black or brown and less frequently yellowish or reddish pigments which are produced in the body through the action of enzymes. Much less common, and not found at all in mammals, are the brightly coloured pigments known as lipochromes. These are fatty materials, soluble in alcohol, which cause the yellow or orange and less frequently the red coloration of animals, for instance the brightly coloured caps of the green, black and spotted wood-peckers.

The cells which contain real pigments are called, in accordance with the various kind of pigment, either melanophores (containing melanins), xanthophores (containing yellow lipochromes) or erythrophores (containing red lipochromes). Some animals, especially fishes and amphibians and more infrequently some reptiles, also have cells called leucophores or iridocytes. These are filled with little crystals of a material called guanine, which is produced in the body as a waste product and which causes the silvery lustre of some animals, particularly fishes. However, guanine can also be deposited in the extra-cellular areas of the skin, sometimes in the internal linings, and is present in the reflective layer of the retina of some animals' eyes.

The pigment cells are generally situated in the skin of animals, either in the lower layer, the dermis, or on the boundary between the dermis and the superficial epidermis. Over and above this, the pigments become incorporated into the products of the skin, whether they be the scales or horny matter of fishes and reptiles, the feathers of birds or the hairs of mammals. Usually, however, only some of the colour pigments penetrate these derivatives of the skin. Only melanophores, for example, occur in the scales of reptiles, even though their skins contain all the different types of colour cells.

In exceptional cases pigments can even occur outside the skin and its derivatives. For example, the greyish colouring of some parts of the central nervous system is caused by melanins.

In some salt-water fishes such as the salt-water garfish there are greenish or bluish pigments mingled with the blood pigment. These are then deposited in various parts of the body and cause green coloration of the bones or a greenish colouring of the dorsal muscles.

Genuine green pigments (from chemicals of the porphyrin group), however, are very rare in animals. They occur in African and Asian touracos, the feathers of which contain so-called touracoid pigments — interesting for the fact that they are fairly readily soluble in water. This is one of the few instances of an animal's colour 'running'.

Even rarer is genuine blue pigment. The blue coloration of animals is generally produced in a manner similar to that which produces the blue colour of the sky or the greenish colouring of thick layers of glass. It is caused by the scattering of light rays in a turbid medium. In animals the 'turbid medium' is cells filled with air or air-filled cavities within the hairs and feathers. These cells themselves reflect a white or greyish colour, but if they have a background of melanophores, the result is the bright blue colouring known, for example, in kingfishers or on the 'wing-mirror' of the jay. It produces the blue coloration in the faces of some species of monkey, the mandrill for instance. If in front of the layer of melanophores there is also a layer of yellow or red pigment, the resulting colour is green or violet. This is what causes the green coloration of the green woodpecker. If instead of the melanophores there is a layer of melaniridocytes, which are iridocytes and melanophores firmly linked together, the result is a blue, green, or violet lustre.

There is yet another way in which the green, blue or reddish lustre is produced in animals, and in particular in birds. This is by means of the phenomenon of interference, produced when light is reflected back from several different levels. The rainbow effect often seen on a patch of oil on a wet road is one common result of the phenomenon. In birds the barbs of the feathers have special lamellae which, through the reflection and interaction of light waves, produce a changeable, almost metallic lustre. This can be seen on starlings and mag-

In reality the pigment cells have a fixed shape. The apparent differences in shape which can be observed under the microscope are caused by the movement of the pigment within the cell. It may be dispersed throughout the cell as in the top picture, or concentrated at its centre. (Drawn after Portmann.)

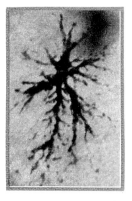

This is how the pigment cells look under the microscope when they are greatly magnified. These cells come from the periphery of the scale.

pies, but is particularly beautiful on hummingbirds. In such cases the apparent colour changes according to the direction from which the light rays are falling. A similar metallic lustre also appears on some mammals. The coat of the otter shrew *(Potamogale)* has a purple lustre, the coat of the golden mole *(Chrysochloris)* has a golden-greenish lustre and the coat of the related *Chalcochloris* has a copper lustre. A silvery or golden lustre is known in rodents such as mole rats *(Tachyoryctes)* and pocket gophers *(Geomys)* and in the marsupial mole *(Notoryctes)*.

Of similar origin to this metallic lustre is the velvet appearance of the feathers of birds of paradise or grebes. It is caused by the fact that microscopic projections occur on the barbs of the feathers.

As already mentioned, the red coloration of animals is produced by red lipochromes. However, the red or pink colouring of the skin can also be caused by the red pigment haemoglobin in the dermis being visible through the surface layers. This is the way in which the pink colouring of the Californian goby fish *(Typhlogobius)* is produced, and it also accounts for the red colour of cockscombs or of the skin on the face of certain monkeys. The red colour of the eyes of albinos is also produced in the same way.

The oddest form of 'coloration' is to be encountered in some colourless fishes. In many species the newly hatched young are transparent simply because the skin pigment has not yet been deposited. The adults of some tropical species of the families Siluridae and Serranidae retain this juvenile characteristic so that with only partially developed chromatophores and transparent skin, together with the small dimensions of their bodies, it is possible for one to see straight through them.

As already mentioned the pigments are either deposited in the skin of animals, or in their products — scales, horny matter, feathers or hairs. If the body covering is thick, as with feathers, fur or even scales and horny matter, the main vehicles of coloration are these surface structures, but the colouring of the individual hairs or feathers is not necessarily uniform. This can be seen clearly in birds, where the pattern of each feather

In the coloration of the domestic cock most of the elements which produce coloration in animals are found. The dark colours are produced by melanins, the bright yellow and reddish colours by lipochromes, the coloration of the wattle is caused by blood capillaries shining through the skin and the metallic lustre of the tail feathers is produced by interference between light-waves reflected from the lamellae of the feathers.

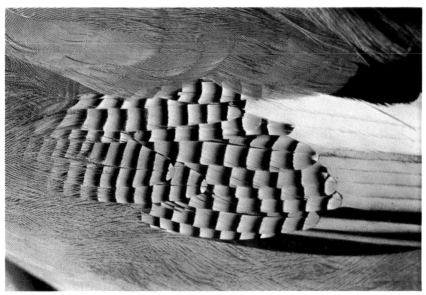

Most people have seen the bright blue 'mirrors' on the wings of the jay (Garrulus glandarius). The blue colour is produced by the scattering of light rays by air-filled cells.

The metallic lustre of feathers is typical of many species of bird. The photograph shows *Lamprocolius abyssinicus.*

Hummingbirds are well known for the bright metallic colours of their feathers. These colours are produced in the majority of cases by light reflected from the delicate lamellae of their feathers, the apparent colour depending on the direction from which the light falls and from which the feathers are viewed. This is clear from the photographs of the ruby-topaz hummingbird *(Chrysolampis mosquitus)*. The two photographs show the same bird, photographed with direct frontal illumination (left) and oblique illumination (right).

There are not many fishes which are 'transparent' throughout their lives. Among those that are is the East Asian fish *Kryptopterus bicirrhis*.

is quite distinct. Furthermore several colour zones can sometimes be distinguished on the hairs of mammals, an obvious example being the spines of the porcupine or the hedgehog.

Each feather of the bird may have its own individual pattern, as in owls or birds of prey. Alternatively the colouring may be just part of the design of a larger unit, as on the wings of ducks and jays or in the overall colouring of divers. In the case of mammals differences in the density of hair in the various zones of the coat may create the pattern, though examples of this are few. The transverse bars of the banded mongoose *(Mungos mungo)* are produced by stripes of denser and sparser hair which to varying degrees reveal the darker lower part of the hairs.

In general birds and mammals have only their bald areas coloured (the head of the condor or the cassowary, the beaks of birds of prey, and the faces and buttocks of baboons), whereas the parts that are covered with hair or feathers are pale, or the pigment only irregularly concentrated. Intense coloration of the skin over the whole body is found in these species only for a transitional period — in mammals, for example, at moulting-time. Polar bears and monkeys are an exception for the skin under their fur is pigmented.

Hairless species of mammal such as hippopotamuses (family Hippopotamidae), elephants (Proboscidea), rhinoceroses (Rhinocerotidae) or whales (Cetacea) have their entire skin pig-

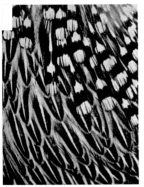

The pigment cells penetrate the products of the skin too and colour the coat of the badger (*Meles meles*), above; the feathers of Sonnerat's jungle fowl *(Gallus sonnerati)*, centre; and the horny matter of the body covering of the European tortoise *(Testudo graeca)*, below.

mented permanently, though anyone familiar with these animals, if only in a zoo, knows that their coloration is not very constant and that it varies within quite considerable limits. This is not caused by the pigment cells, or even by the structure of the skin or hairs, but by external factors which can, even if temporarily, also influence the colouring of these animals. With elephants and rhinoceroses such a factor is, for instance, the colour of the mud that dries on them, or the colour of the sand which elephants throw over themselves. In the same way some whales generally have a brownish yellow colour all over their bodies from a thin deposit of algae. Another factor may be the blood from hunted animals which sticks on the hair of beasts of prey and which can give rise to a rust-coloured tint over a period of time.

The coats of many animals are coloured by secretions from various glands. For example, there is a pink to red coloration on the breasts of some male kangaroos (Macropodidae) and in wintertime the stoat *(Mustela erminea)* has a yellowish colouring to the underside of its coat. The thick-tailed opossum *(Lutreolina crassicaudata)* sometimes has a pink hue to its coat which on the underside of the body deepens to a brick red shade, but which disappears soon after the death of the animal. It is assumed that the cause of this colouring is also a secretion.

Finally, to demonstrate all the factors that can influence the coloration of animals, mention must be made of sloths (Bradypodidae), which in their natural habitat have a greenish colour that quickly disappears if the animal is killed or even kept in captivity. This greenish coloration is caused by symbiotic algae which live on the coats of sloths in the layer under the cuticle of the hair. There are several types of green and bluish green algae which are often typical of the individual host species. Algae are also probably the cause of the greenish colouring of the bare-backed fruit bat *(Dobsonia peroni)*.

Even this brief review shows that the origin of animal coloration is a relatively complex matter and that many factors are involved. However, more interesting still is the result of these processes, animal coloration itself in all its richness and diversity.

COLOURS FOR RECOGNITION

In the book *Two Little Savages* by the American author and naturalist Ernest Thompson Seton, one of the heroes tells the story of a boy who was learning to identify animals. One day the boy saw a wild duck on the lake. It was too far away for him to recognize its colouring and all he could see were some coloured spots. He made a drawing to show the position of these spots on the body of the bird and later he compared his drawing with pictures of ducks in a book he had at home. He soon found the kind of duck he had seen, for the spots he had drawn had been sufficient to identify it. He realized that different kinds of duck have different spots and stripes which distinguish them from one another. All one has to do is remember these simple 'uniforms' in order to recognize their wearers, even at a great distance.

The pattern in the coloration of birds may be produced in two ways. Either each feather has its own special relatively independent pattern, as on the feathers on the underparts of the Eurasian sparrowhawk (*Accipiter nisus*), on the left, or the coloration of the individual feathers is just a part of a larger unit, as with the feathers forming the wing 'mirror' of the common teal (*Anas crecca*), on the right.

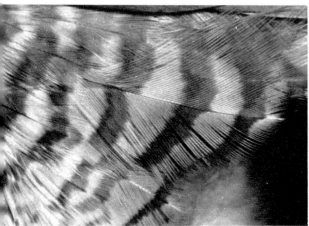
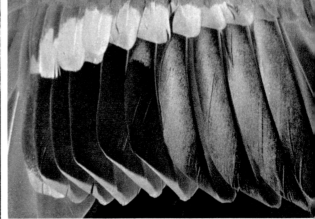

The positions of the main colour patches in several species of duck. In the upper row (from left to right) are the common teal *(Anas crecca)*, the northern shoveler *(Anas clypeata)*, the mallard *(Anas platyrhynchos)* and the pintail *(Anas acuta)*. In the lower row are the tufted duck *(Aythya fuligula)*, the pochard *(Aythya ferina)*, the goldeneye *(Bucephala clangula)* and the red-crested pochard *(Netta rufina)*. (The exact shape of the body and the relative dimensions have not been reproduced.)

This is no doubt a little hard to grasp for the city dweller, who generally only sees animals in a museum or a zoo. After all, he is used to seeing animals close up and to noticing all the details of their coloration. He usually notices a large number of details which, though interesting, are of little value in recognizing the animal and which are also difficult to remember. But out in the wild, where animals are normally seen only from a distance, the details disappear, the halftones are lost and an overall impression is created by a few characteristic spots and the contrasting colours — generally black and white — because most of the other colours either merge with them at a greater distance, or create an indefinite, uniform grey. It is this simplification which produces the animals' 'identikit' pictures which can then be easily remembered, making possible quick identification by human beings. For the time being animal coloration will be considered only from our human point of view.

The attractive appearance of ducks makes them a very rewarding group for practising identification by colouring, but similar characteristics can be found in many other animals. For example, there is the dark zigzag line on the back of the adder *(Vipera berus)*, the yellowish white collar behind the head of the grass snake *(Natrix natrix)*, the white half-moon patch on the breast of the ring ouzel *(Turdus torquatus)* or the black and yellow colouring of the spotted salamander *(Salamandra salamandra)*. There is also the distinctive black and white colouring of the giant panda *(Ailuropoda melanoleuca)*, the black stripes of the tiger *(Panthera tigris)* or the colouring of the caparisoned Malayan tapir *(Tapirus indicus)*.

However, the more animals there are from which to select typical colour characteristics, the greater is the probability that two species will occur that are the same or very similar in their coloration. This is the case with some common species of frog. The edible frog *(Rana esculenta)* and the marsh frog *(Rana ridibunda)* differ from one another only in the colour of their vocal sacs. The pine marten *(Martes martes)* and the beech marten *(Martes foina)* are alike except in the shape and, less reliably, in the coloration of the throat patch. It is difficult to separate the various species of spotted woodpecker and almost impossible to differentiate between the yellow-necked field mouse *(Apodemus flavicollis)* and the long-tailed field mouse *(Apodemus sylvaticus)*.

In some cases, geographical distribution can help with identification. The European mink *(Lutreola lutreola)* is on the whole quite easy to recognize out in the wild if it is known that nothing else with a similar colouring lives in the area in question, but if it is seen in a zoo alongside an American mink *(Lutreola vison)* it is much more difficult to distinguish between them. Similarly it is hard for the layman to differentiate be-

One of the most characteristically coloured vertebrates is the giant panda *(Ailuropoda melanoleuca)*, a rare beast of prey which lives in the province of Szechwan in China. The photograph shows the female Chi-Chi from the London Zoological Gardens.

The pine marten (left) and the beech marten (right) can best be distinguished from one another by the shape of the throat patch, although even that is not absolutely constant.

tween the leopard *(Panthera pardus)* and the jaguar *(Panthera onca)* or between the various species of gulls, if geographical distribution is not taken into account. This is only of partial help in separating species such as the great black-backed gull *(Larus marinus)* and the lesser black-backed gull *(Larus fuscus)* which have an overlapping distribution. Even though the two species differ markedly in size, this is not always obvious in the field and where the two kinds occur together a difference in leg coloration is the main differentiating characteristic, and this is not at all clear at a distance. And yet one should not claim that such species cannot be distinguished according to their colouring; on the contrary, there are many species of fish that can only be identified by coloration, as all their other characteristics are identical. These differences are not in general all that striking and usually escape the attention of the layman. The difference between the leopard and the jaguar, for example, is that the jaguar usually has inside its dark blotches one or several more black spots, whereas with the leopard these blotches are empty.

Not only species that are closely related have similar colouring. By an evolutionary process known as convergence, almost identical coloration can sometimes be produced in species that are very distant relatives and which are also geographically distant. For example, not only two species of the marten family, the American striped skunk *(Mephitis mephitis)* and the African polecat *(Ictonyx striatus)*, but also the Australian striped phalanger *(Dactylopsila trivirgata)*, have a distinct black and white colouring on their coats.

In nature nothing is usually simple and this is true of animals' colour patterns too. So far it has been assumed that the colouring of animals is fixed and that all the individuals of one species are coloured in the same way. This assumption is not, however, completely correct. Just as no two human beings are quite the same, neither are any two animals. They can differ in size and in many other physical characteristics, among which is coloration. Of course, this does not mean seasonal or age or sex differences, but merely those differences which occur in individual animals regardless of their age, sex or geographical distribution. Naturally these differences vary in extent. In some species there are on the whole only minor deviations, for example, in the pattern, which escape the observer on a cursory viewing. But in other species the differences are such that they will be noticed even by a layman. Ash grey female cuckoos appear side by side with brown ones; rust-coloured red squirrels occur together with black ones. These differences in coloration are called colour phases. Animals coloured in this way occur — unlike geographical races — next to one another in the same territory. It is interesting to note that sometimes one of these phases may gain such a degree of predominance that in a certain territory it achieves a frequency of a hundred per cent. For example, on the European continent red squirrels occur in the rust-coloured and black phases, while in England there are only rust-coloured squirrels.

Two-colour phases occur relatively frequently. Apart from the above-mentioned examples there are two-colour phases in many beasts of prey of the cat family, for instance in pumas *(Puma concolor)* or in jaguarundi cats *(Herpailurus yagouaroundi)*, which are known in greyish, greyish blue and reddish brown colour phases. In some species, however, there are many more of these colour phases. For instance, North American black bears *(Ursus americanus)* are known, in addition to the basic black colour, in brown, cinnamon, almost white and in a special light bluish colour (so-called glacier bears). Among marsupials and monkeys one troop may contain animals with many shades of colour, sometimes in such numbers that it is really no longer possible to speak of individual colouring. Some species are known where practically every animal has its own individual colour or type of pattern.

Today there are three species of zebra in the world. Though they look almost the same at first glance, they can be distinguished by the type of striping, especially on the rear part of the back. The various species are illustrated in this picture: above *Equus grevyi*, centre *Equus zebra*, below *Equus burchelli*. (Drawing modified after A. Rzasnicki.)

Among the animals re-nowned for their great colour variability is the fighting cock *(Philo-machus pugnax)*. Differ-ences in the coloration of males are so great that it is difficult in one flock to find two birds coloured in exactly the same way.

Renowned for this are fighting-cocks *(Philomachus pugnax)* and wild dogs *(Lycaon pictus)*, of which it is claimed that one cannot find two identically coloured animals. Also whales (Cetacea), for example the common killer whale *(Orcinus orca)* or some dolphins, may show great differences in the coloration of individual animals. In the case of the South American spec-tacled bear *(Tremarctos ornatus)* the variability is limited to the face pattern. The face mask, however, is fixed throughout the life of each individual animal, just as it was formed in the young cub.

In spite of all these more complex examples, the main impact made by an animal on a human being is caused by its colour patterns providing a characteristic series of visual symbols for each species which can then be quickly and easily recognized. The spectacled bear can be identified whatever form its face mask may be, because there is no other bear with a similar coloration. In order to recognize the skin of the jaguar one single spot from its pattern is sufficient; an adder can be identified even in a caricature of a snake if it has a zigzag band on its back. However, there must be absolute

certainty as to the coloration of the animal in all the forms and variations that occur in the respective species.

Before studying in more detail the differences and variations to which animal coloration is subject and thus complicate the question of coloration even more, it is necessary to try to find in this great and apparently chaotic wealth of colours and patterns, some principles, some common plan, that will meet the human need to classify everything and arrange it in groups with common attributes.

PATTERNS AND COLOURS

The South American spectacled bear *(Tremarctos ornatus)* is the only bear to have a characteristic light pattern on its face. The shape varies considerably from animal to animal. (Drawing modified after Roth.)

Defining the various types of coloration which are of general validity, at least in vertebrates, is a very difficult task. Consider the multiplicity of coloration in fishes and the vast number of incredible combinations of colour and pattern. The coloration of other vertebrates is much more subdued in comparison with that of fishes and yet even there one finds a surprising variety of colours.

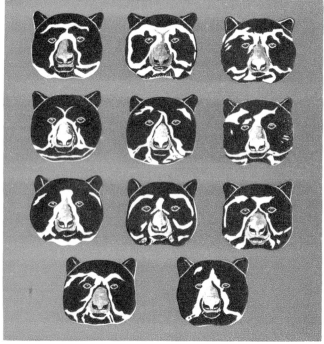

The main concern of this chapter is to define and categorize animal coloration by means of fairly superficial characteristics without considering any more profound relationships of colour and pattern. Grouping in this way will provide an ordered framework for a systematic study of the great variety of forms and colours.

The first group consists of single-coloured animals, including those which have a lighter underside to their bodies or those which have only one smallish coloured patch in their coloration, for example, the red cap on the head of the black woodpecker *(Dryocopus martius)* and the white pattern on the chest of the Himalayan bear *(Selenarctos thibetanus)* and the Malayan sun bear *(Helarctos malayanus)*.

The coloration of animals of the second group consists of relatively small spots or patches such as appear on leopards, giraffes and many young gulls.

The third group includes animals whose coloration consists of bars or stripes across or along the body as on the tiger, the zebra, and some kangaroos.

The coloration of animals in the fourth group is formed either completely or mainly by a small pattern duplicated over and over again. Tortoises, cuckoos, birds of prey, some owls and many snakes are examples of this group.

The final group includes all the remaining types of pattern. Under it fall animals whose colouring is formed by large irregular patches or colour areas as in the wild dog *(Lycaon pictus)*, parrots, and the European salamander *(Salamandra salamandra)*.

The coloration of young black-headed gulls *(Larus ridibundus)* is highly variable and in individual birds the size, shape and density of the dark spots varies greatly. The photographs show chicks aged thirteen to fifteen days.

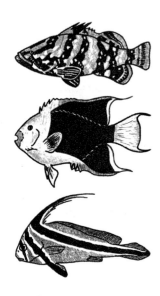

The variety in the coloration of fishes is enormous, both in colour and pattern. It is impossible to describe and very difficult to define general characteristic types. Thus the picture merely shows the multiplicity of pattern of some fishes. From top to bottom are illustrated the patterns of fishes of the genera *Epinephelus, Holacanthus, Eques* (p. 30), *Arothron, Pterophryne* and *Pomacanthus* (p. 31). The fishes are not drawn to scale. (Drawings modified after various sources.)

With the above groupings in mind it is now possible to see whether there are any closer relationships between the various types of colouring. It is known, for example, that the arrangement of some conspicuous elements of the pattern is not constant in all vertebrates. In the lower classes, such as fishes and amphibians, these elements are positioned quite arbitrarily on a fin or on the tail for example — whereas in birds and mammals they normally occur at the ends of the body, particularly on the head and to a lesser extent on the posterior regions. The same general principle seems to operate *within* the various classes of vertebrates with more advanced members demonstrating a greater precision in the arrangement of patterns. To some extent, therefore, it appears that the type of colour arrangement is associated with position on the evolutionary scale.

These relationships can be illustrated using the mammals as an example. Their colouring is not as varied as, for example, that of birds and the distinctiveness of their appearance is not produced by the colours so much as by the patterns. However, not even the patterns of mammals are as varied as those of some other groups of vertebrates.

For the time being this comparison will not include those species with large patches of irregular shape, such as the common rorqual *(Balaenoptera physalus)* with its asymmetrical colouring of the lower jaw, white on the right half and grey on the left.

Thus three basic types of coloration remain: monochrome, spotted and striped. The most advanced of these is now considered to be monochrome, primarily because in the development of the coloration of some species there is a trend to a single colour, with the gradual disappearance of stripes and spots. Evidence of this can be adduced from the coloration of the exterminated quagga, which was striped only on part of its body, while the rest was an overall brown colour. It can also be seen in the coloration of some antelopes, for instance the sitatunga *(Limnotragus spekei)*, in the male of which the pattern is generally completely lost, while the young and the female are still striped.

Monochrome coloration is also considered developmentally more advanced because many species of animal are single-coloured in adulthood, but have striped or spotted young, or because the stripes are at least partly lost in adult animals. For example, young Burchell's zebras are densely striped even low down on their legs, whereas the old males of this species have almost pure white legs.

There are many cases where in the young animals there appears a more primitive striped or spotted effect, while the adult animals are monochrome. Sometimes it also happens that such coloration, or a part of it, remains on the animal when it grows to adulthood. The four-year-old lion on page 34 still had a clearly recognizable pattern of spots and stripes. This lasting of the pattern of young animals into adulthood is the third reason why monochrome colouring is considered to be a more advanced type of coloration.

In some single-coloured or almost single-coloured animals remnants of the earlier pattern can be found even today. The 'ass's stripe' is considered such a remnant of a striped pattern. It is found not only in asses and some other odd-toed ungulates, but also in many antelopes and deer. With it can be compared the dark band on the back of the field mouse *(Apodemus agrarius)* and of the northern birch mouse *(Sicista betulina)*. The two-toed anteater *(Cyclopes didactylus)* has a similar dark stripe on its back when it is young, but in adulthood this colouring is lost. The brown hare *(Lepus europaeus)* has a dorsal stripe that is to be seen only in the embryo. Some zoologists are of the opinion that these stripes are relics of a dorsal or abdominal mane which the nyala antelope *(Tragelaphus angasi)* has to this day. The original pattern is preserved for a relatively long time on the shoulders and legs and can be observed, for example, on some Przewalski horses.

The remaining two types of coloration, spots and stripes, are both considered to be very primitive characteristics and are frequently found in forms lower on the evolutionary scale. Many marsupials are patterned in this way, one example being the striped Tasmanian wolf *(Thylacinus cynocephalus)*. In this connection it is interesting to compare these primitive forms of mammal with members of evolutionarily higher orders in

The coloration of the snowy owl *(Nyctea scandiaca)* is an example of a pattern created by elements occurring again and again.

which both spots and stripes are also often found. In the primitive forms the stripes effect only the rear half of the body, whereas the area of the head is usually single-coloured, without any emphasis of the structure. In the higher mammals, however, the pattern is usually more complex on the head, even in cases where the rest of the body has practically no pattern at all, as in the coloration of the ears of the dwarf forest buffalo *(Syncerus caffer nanus)* and the Malayan sambar *(Rusa equina)*. The same is true within individual orders. For example, primitive ruminants like chevrotains (Tragulidae) have a pattern of stripes or spots on their bodies which hardly emphasizes their heads at all. On the other hand, the ruminants highest in the evolutionary scale such as antelopes and aurochs generally have conspicuous patterns on their heads.

The mutual relationship between these two types of pattern is interesting. Even though there is no complete agreement of opinion in this respect, it would seem that stripes are more primordial and that spots may arise when the stripes disintegrate. The fact that spots are very often clearly arranged in rows along the body bears out this point. The white stripes along the body of the paca or spotted cavy *(Cuniculus paca)* and young tapirs, for instance, often break down into spots.

There are basically two types of stripes, transverse and longitudinal. It would not seem, however, that the direction in which they are aligned depends on the position of any of the internal organs, as was at one time assumed. The stripes either run regularly over the whole body, or the two systems interfere with one another, as on the haunches of zebras. In other animals the direction of the stripes may differ on the body and on the limbs. These differences can generally be explained by the position of the limbs in the embryo and the state of the pattern at the moment the limbs are drawn away from the body, as illustrated in the drawing on page 39.

At the places where there is interference of the stripes,

Coloration consisting of two large colour areas is shown on this interesting breed of domestic goat from Switzerland — the Valais goat.

33

a disintegration into a system of spots often occurs which sometimes actually enables a fluid transition of the horizontal stripes on the limbs into the vertical stripes on the body. As has already been mentioned, it is probably this disintegration of the stripes which actually produces the spotted effect. Evidence for this is provided by the finding of several abnormally coloured animals which show the manner in which the spotted coloration probably originated. One such find is linked with the interesting story of the description of a new species of cheetah. All present day cheetahs *(Acinonyx jubatus)* are relatively densely and irregularly spotted, but in Asia an animal was shot which was completely covered with stripes along the body and it was described as a new species of cheetah *(Acinonyx rex)*. However, as only this one single example is known, it is obvious that it was an abnormal colour variant of the common species, particularly as a similar colour aberration had already been observed in the leopard *(Panthera pardus)* and the serval *(Leptailurus serval)*. In the latter the

The coloration of this four-year-old lion *(Panthera leo)* has retained the juvenile pattern of striking, almost continuous stripes.

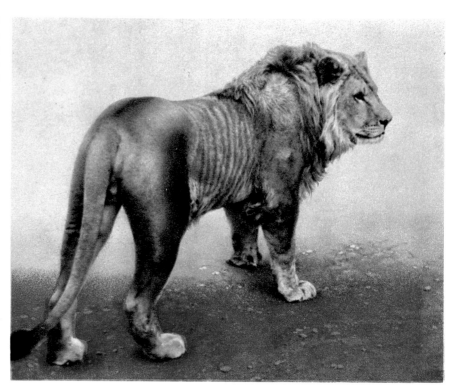

The coloration of the ornamental upper tail-coverts above the tail of the peacock *(Pavo cristatus)* is an example of a pattern produced by the repetition of the separate patterns on individual feathers.

In birds, generally only exposed areas of skin are pigmented. In this photograph of the king vulture (*Sarcorhamphus papa*), the naked but brightly coloured head is a conspicuous feature.

Bright, striking colours produced by lipochromes are typical of bird coloration. The photograph shows *Pyromelana granciscina*.

The 'ass's stripe' on the back of mammals is sometimes considered a remnant of the dorsal mane. This mane is still known today in several species of antelopes (see the photograph on p. 70). It would also appear that in this picture of an Asiatic wild ass *(Equus hemionus)* the dark-coloured hair forming the dorsal stripe is longer than the hair on the rest of the body.

spots mingled with the longitudinal stripes, leaving only small spaces in between.

Incidentally in some species of mammal, one can observe both types of coloration quite commonly. For example, the above-mentioned serval has relatively dense black spots all over its body which link up with the black stripes along the neck.

This somewhat mechanical evolutionary view of animal coloration is in itself too one-sided, as it does not take into account the other factors which no doubt exist, such as adaptation to various habitats, colour change, the signal function of coloration and a number of other factors. But it would be wrong to omit this more or less descriptive approach completely, as it helps to give an idea of the origin of coloration and of the mutual relationships between the various elements.

COLOUR CHANGES

The ability of animals to change their colour has been known to man since the time of Aristotle and it is still one of the most favoured topics when animal coloration is under consideration. It has even managed to make chameleons generally known and popular, though not with full justification, since they are not the only, or the most typical group of animals that are masters of this art. The ability to change colour, whether for a short time or for a longer period, is much more widespread in the world of animals than is generally realized and there are also a large number of ways in which colour changes are produced.

The almost extinct Tasmanian wolf (*Thylacinus cynocephalus*) is a mammal which retains a typical primitive striping on the rear parts of the body.

As well as the chameleon, many other animals can change their colour to a considerable extent. For instance, among the iguanas, the Carolina anole *(Anolis carolinensis)* can change its colour from a dark brown to a light green. Among the amphibians the tree frog *Hyla goughi* from Trinidad is apparently without competitors in this field, for its colour scale ranges from dark brown by way of a reddish brown and various shades of yellow to a very pale greyish white. This manner of colour change is even more common in fishes. The American zoologist C. H. Townsend, for instance, when moving fishes from the region of the Bermudas to an aquarium in New York, observed sudden changes of colour in twenty-eight species of fishes of which some, for example the Nassau grouper *(Epinephelus striatus)*, were able to alternate between up to eight different liveries within a few seconds. In some sharks and rays (Selachii) too, there occur considerable changes in colour when the environment is altered. An American research worker, G. H. Parker, carried out experiments with pairs of

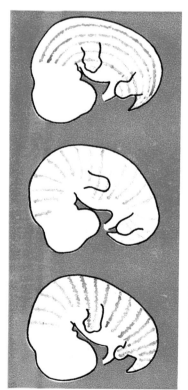 

This diagram illustrates the relationship between the position of the limbs in the embryo and the direction of the stripes in the adult. Above, the type of striping of a young tapir (stripes lengthwise all over the body), in the middle that of the striped hyena (vertical on the body, horizontal stripes on the legs) and below, that of antelopes (vertical stripes all over the body). (Drawing modified after Krumbiegel.)

In the damar zebra (*Equus burchelli antiquorum*) the pattern on the thighs breaks up into irregular spots.

In the damar zebra (*Equus burchelli antiquorum*) the pattern on the thighs breaks up into irregular spots.

skates *(Raja erinacea)* of the same colour. He put one into a tank with the walls painted a light colour and he placed another in a tank with dark walls. After eighteen hours the first fish was light cinnamon-coloured, while the other was very dark. When he moved the fishes to the opposite tank, the light one became dark in nine hours and the dark one changed its colour in twenty-one hours to a pinkish white.

In the wild such changes are sometimes caused in reef fishes when they swim up from the bottom to the surface. The greatest change of this kind is known in the great surgeon-fish *(Hepatus matoides)*. These fishes are coloured black, with the exception of the tails and light fins, when moving around on the seabed, but as soon as they rise eight to ten metres from the bottom they change to a greyish blue colour. Another fish of the genus *Thalassoma* is yellow when moving around near the seabed and dark blue when in open water. For a long time it was described as two different species, the blue *Thalassoma nitida* and the yellow *Thalassoma nitidissima*.

Change of colour can be initiated by visual stimuli as in the case of blind fishes which lose the ability to change colour and are mostly dark, or by environmental factors such as light, warmth and moisture as with the tree frog. Internal factors such as the level of the sex hormones and secretions of the adrenal and pituitary glands for example, together with the state of the nervous system, also have an influence on colour changes. The actual change of colour is made possible by the fact that the chromatophores (with the exception of the iridocytes) are linked up to the nerve endings which direct the movement of the pigment. When at rest the pigment is usually freely dispersed through the chromatophores. When the cell is stimulated the pigment becomes concentrated at the centre. Such a change of colour, common in fishes or chameleons for instance, takes place very quickly — much more quickly than in amphibians or other species of reptile where it is regulated by hormones.

The influence of visual stimuli on colour change is best

The pattern on the coat of the serval *(Leptailurus serval)* clearly shows the transition from the stripes along the neck into the spots on the body. The photograph also illustrates how the spotted coat makes it more difficult to identify the outline of the animal's body.

The best-known animals capable of colour change are chameleons. The photographs show one and the same animal of the species *Chamaeleo chamaeleon* at rest (left) and when irritated (right). In the picture on the right one can also see the slightly opened mouth of the irritated animal. In both pictures the eye is camouflaged by the linked lids, covered with scales, and with a pattern which makes them appear continuous with the surrounding area of the head.

seen in the plaice *(Pleuronectes platessa)*, a flatfish the coloration of which is controlled by the background colour in the range of vision of the fish's eye. If the flatfish is lying on the border of two contrasting colours so that head and eyes are on one side and the rest of the body on the other side of the borderline, the whole body has the colour of the environment in the vicinity of the head, regardless of whether the larger part is lying in the other colour area.

Also of interest are the sudden changes of colour in some fishes which take place as a result of external stimuli. Sometimes the colouring differs not only according to the various situations, but also according to the sex of fishes in the same situation, as in the illustrated species *Apistogramma reitzigi*. Similar changes occur in other animals too. Chameleons change their colour as a result of visual stimuli, and also as a result of fright, anger, cold, warmth or irradiation.

The actual colour change, which is often very impressive, is brought about in a relatively simple manner — by dispersing or concentrating the brownish black and yellow pigments. Their mutual combination, in accordance with the principles described in the first chapter, then produces the most diverse coloration, from brownish black to bluish green or yellowish green.

This manner of colour change, however, is only feasible in reptiles, amphibians and fishes, where the pigment is deposited in the chromatophores and can move around in them. In mammals and birds, where the pigments are deposited in the dead products of the skin (hair, feathers) or in the melanocytes (skin cells in which there is no movement of the pigment), such a change is not possible and minor sudden changes of colour can only take place on bald spots through a change in the intensity of the blood supply. The uakari monkey *(Cacajao rubicundus)* goes red in the face when it is excited, the ears of the Tasmanian devil *(Sacrophilus harrisii)* also turn red and the turkey's wattle changes colour.

Yet there is a way which enables these animals to make extensive changes in their coloration. It is not caused by movement of the pigment, however, but by movement or physical wearing down of the hair or feathers. In the chapters on warning coloration and inter-communication between animals more details will be given about various visual signals

Though it is not generally known, of all vertebrates fishes have the greatest ability to change their colours. The photograph shows two fishes of the species *Botia lohachata*. The difference in coloration is not caused by variability (which is a permanent deviation in coloration), but by a sudden change of colour.

A detail of the feathers of a starling *(Sturnus vulgaris)*. Starlings have this coloration in autumn after moulting. By spring the white tips of the feathers are worn off and the starlings are then dark, almost black. However, the metallic lustre which can be seen clearly in the photograph is permanent.

in mammals and birds that consist of a sudden change in the coloration of a certain part of the body. For instance, the showing of the white 'star' in deer is caused by the white hairs, which when at rest lie flush with the body and are mostly covered by the surrounding brown hair, but which bristle up when the animal is excited, thus revealing their colour. In a similar manner the springbok *(Antidorcas marsupialis)* when excited shows a broad band of white hairs which when at rest lie in a fold of the skin on the back and are almost completely covered by the surrounding dark hairs.

Another principle of colour change involves two or more colours in hair or feathers. When they are at rest, lying flush with the body, the base parts are covered and only one colour can be seen. The lower, differently coloured parts only come into their own when the animal is excited and the feathers or hairs 'stand on end'.

In some animals, there may occur over a longer period of time a colour change caused by a physical wearing out of their outer covering, for example in birds by the wearing down of their feathers. In the autumn, starlings *(Sturnus vulgaris)* have dark feathers with white tips, so that they look speckled. In spring, however, the white feather-tips are worn down so that the white in their coloration disappears and they become darker. Similarly, male chaffinches *(Fringilla coelebs)* have a browner head after their autumn moult. By spring, however, the brown ends of the feathers wear down and the colour of the head changes to a bluish grey. Still, there can only be a fundamental change in the colour of mammals and birds when they moult or shed their feathers.

Finally mention should be made of a change which cannot really be called a genuine colour change at all, since it takes place after the animal has died. Such changes in coloration are apparently caused either by the decay of the nerve or hormone control mechanisms or simply by the breakdown of the unstable fatty pigments. These changes are best known in fishes. Even in ancient Roman times the red mullet *(Mullus barbatus)* which lives in the Black Sea and the Mediterranean, was carried alive to the table so that the guests could watch the gradual changes in the coloration of the dying fish.

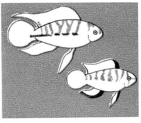

This diagram of the colour changes in the fish *Apistogramma reitzigi* shows how fishes can react to various stimuli with a change in colour. At the same time one can see that the colour reactions of the male and female to the same stimulus are different. The drawing illustrates the reaction of the male and female in three different situations: above, threatening coloration, centre, mating coloration and below, alarm coloration. The male is larger than the female. (Modified after Vogt.)

45

ANOMALIES IN COLORATION

Albino tiger cubs *(Panthera tigris)* in the Bristol Zoo. The stripes are black with a shade of brown, the eyes are light blue in the region of the pupil, yellowish at the edges.

Albino tigers are kept only in very few zoological gardens. They all come from Central India, from the stock of the Maharaja of Rewa. Their story is interesting. Over the last fifty years nine albino tigers have been observed in the former state of Rewa. The last of these, a nine-month-old male, was caught in 1951. When he grew to adulthood, he was mated with a normally coloured tigress, also caught in the wild. In three litters a total of ten cubs were born, but all of them were normally coloured. One of these normally coloured daughters was mated with her albino father as a result of which four albino cubs were born (three females and one male) in the first litter and three cubs (of these, two were albino males) in the second litter. Some of these cubs were sent to zoological gardens in the USA and Europe and became the basis for the present stock of 'white' tigers in these zoos.

Just as the ability to change colour arouses people's interest, so do colour deviations in the appearance of animals, especially those that differ markedly from the normal colouring of their species. Considering the singular effects produced by the extreme cases of deviant coloration — albinism (completely white colouring) and melanism (completely black colouring), one should not be surprised that they attract so much human attention. The white colouring of an animal, when encountered in nature, can have an almost supernatural effect. It is no coincidence that it has often found a place in works of fiction and that a number of these animals occur in folk stories

and legends. For the peoples of the North the white reindeer is such an animal, and one of the most famous albinos of literature is Herman Melville's white whale 'Moby Dick'. It is interesting to note that white whales are actually known, for example a semi-tame solitary albino dolphin *Grampus griseus* was known as 'Pylorus Jack' to the crews of all ships navigating Cook's Passage.

Colour variants are not as rare as is usually thought and they are not limited to albinism and melanism. It is possible that various colour anomalies occur in all, or most species. The fact that we do not know of all these variations is a result of the

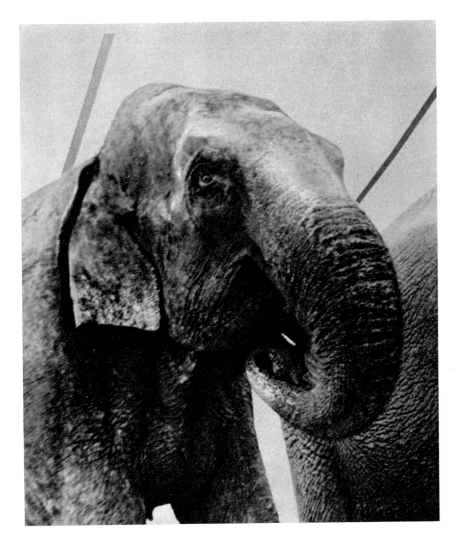

Indian elephants *(Elephas maximus)* often lose the pigment in their ear lobes and at the roots of their trunks. These areas are then light pink in colour.

Melanistic adders occur
mainly in mountainous
regions.

small numbers in which some species occur or their retiring manner of life. Albinos have been observed in well over half of the species of mammals living in England. Among 414 jaguar skins of diverse origin nineteen black individuals were found. Similarly among eighty-seven skins of the South American tayra *(Eira barbara)* of the stoat family there were eleven black skins. African civets *(Civettictis civetta)* are up to forty-five per cent black, and in the region from Bali to Lombok the palm civet *(Paradoxurus hermaphroditus)* occurs only in the black form.

Apart from albinism and melanism there are a number of other colour deviations, which differ from them basically only in their intensity. Albinism, for instance, is an extreme case of hypochroism, which is coloration caused by a fall in the amount of pigment. Albinism occurs when the pigment disappears completely and then the whole animal is coloured white. Genuine albinos have red eyes because the pigment is lacking in the iris too.

The opposite of hypochroism is hyperchroism, which is

coloration caused by an increase in the amount of pigment. Animals coloured in this way are very dark and, in extreme cases, completely black. The latter are melanistic individuals. Just as not all white coloration is albino, melanism should not be confused with the natural black colouring of animals such as the Himalayan bear *(Selenarctos thibetanus)*, the crow *(Corvus corone)*, the binturong *(Arctictis binturong)*, which is a relative of the palm civet, and many others.

Melanism usually affects the whole body and is only rarely partial. On the other hand partial albinos are relatively common, occurring in many familiar species including the blackbird *(Turdus merula)*. In this species quite large areas of the body may be speckled white or there may simply be a number of tiny white spots. An example of partial albinism is the white colouring of the tips of mammals' tails, which has been recorded in the majority of species living in England.

There is a difference between partial albinism and schizochroism, which occurs only when some of the pigments are lacking. The latter causes various partial coloration effects, such as light brown blackbirds and white goldfinches *(Carduelis carduelis)* with red face masks. The most common cases of such anomalies are flavism, the light yellowish brown colouring of otherwise dark animals, rufinism, the abnormal reddish brown colouring of dry-land vertebrates, and the related rutilism, the red colouring of fishes. Similarly, through the absence of all pigments except yellow (xanthorism), the 'golden' forms of some fishes are produced.

Speckled blackbirds *(Turdus merula)* are a relatively common phenomenon and are frequently encountered in city parks. They are partial albinos which to varying degrees have lost their pigment — from tiny spots to extensive areas among which only little remains of the original black coloration.

Other less common deviations are phaeomelanism and chlorism. Bluish coloration (cyanism) is very rare and is caused by the coming together of a considerable number of factors; it is known, for example, in wolves *(Canis lupus)*, minks *(Lutreola vison)*, skunks *(Mephitis mephitis)* and otters *(Lutra lutra)*. Some animals, for instance the 'blue' black bears (so-called glacier bears) are very rare and much sought after. In other species the occurrence of 'blue' creatures is irregular and depends on the region in which they live. For example, 'blue' arctic foxes *(Alopex lagopus)* form only three or four per cent of the total catch in Canada, and in Alaska one or two blue examples are found in every 1000 skins. On the other hand, in Greenland, as is clear from the detailed statistics that have been kept for more than twenty-five years, the blue phase of the fox is predominant.

Of interest is another colour anomaly, which seems to occur only in fishes. In this case all the pigment cells are present, but the iridocytes are missing. The fish are therefore coloured, but they have no lustre.

A rufinous woodcock *(Scolopax rusticola)* (left) in comparison with the normally coloured bird (right). In this case only the darkest element of the pattern is missing. Note particularly the largish dark spots at the back of the neck and on the back of the normally coloured bird.

Colour variations, however, are not merely of academic interest. They have serious consequences for many animals and also raise a number of other questions. It is interesting to note, for example, that in some species colour variations occur relatively frequently whereas in others they are extremely rare. One can quote as an example of this the relatively frequent occurrence of albinos among wolves which live north of latitude

A young albino gorilla *(Gorilla gorilla)*, called Copo di Nieve (Snowflake), in the Barcelona Zoological Gardens. It was caught wild and the mother was normally coloured.

40°N. Among 200 skins of the tayra there were twenty-four partial albinos and one complete albino. About five per cent albinos are recorded in skunks. Of 36 000 moles *(Talpa europaea)* from the region of Modena fourteen animals were white, six grey, two yellowish and one partly yellowish red, partly grey.

On the other hand, there are species in which colour variants are extremely rare. A pure white or a completely black beaver *(Castor fiber)* is a rarity, for example. The proportion of albino bison to normally coloured animals is recorded as one in five million. Albino elephants are very rare, too, though examples are found with very pale skins and with white patches on their ears. So-called 'white elephants', however, are mostly coloured artificially. In some species colour anomalies have so far not been recorded at all. For instance, no one has reported finding a single abnormally coloured individual among thousands of fruit-eating bats or among toads, and colour variants appear to be exceedingly rare in lemmings *(Lemmus lemmus)*.

However, the percentage of abnormally coloured animals in a given species is not necessarily constant. According to information from literature, for example, over the last 150 years there has been an increase in the number and distribution of melanistic individuals of the European hamster *(Cricetus cricetus)*, and a similar phenomenon has been observed in the rabbit *(Oryctolagus cuniculus)* over the last fifty years in Tasmania. On the other hand, the occurrence of the black form of the fox in Northern America has decreased greatly over the last century.

Melanism is at present considered to be the most common colour abnormality. It is probable that it is determined by the dominant gene and it is to a certain extent related to environmental conditions. Albinism, on the other hand, is generally considered to be a pathological phenomenon which is a manifestation of degeneration and which may occur together with other disorders. Its frequency is perhaps influenced by the degree of inbreeding which has taken place and this in turn may be affected by a species living in very close proximity in

The small parrot *(Trichoglossus haematodus)* is one of the most brightly coloured birds.

The monochrome colouring of the fennec *(Fennecus zerda)* is a typical example of adaptation to desert conditions.

Light patches on a dark background are a common element in the cryptic coloration of animals living in forest and wooded regions. The photograph shows the male of the bushbuck *(Tragelaphus scriptus)*.

colonies or in captivity. It is suggested, for instance, that albino rodents occur especially frequently in periods of over-population. In the 'mouse years' many spotted individuals or animals with a white stripe occur among fieldmice, even though complete albinos are relatively rare.

It is interesting to note that whereas albinos are very often weak and sickly, melanistic animals are frequently stronger than their normally coloured relatives. It is affirmed, for example, that the black racoon can be as much as thirty per cent heavier than a normally coloured animal of the same species.

Although it is generally believed that abnormally coloured animals are driven away or killed by other members of the species, and observations have been made of albinos which kept their distance from other normally coloured animals and often stayed in the vicinity of human settlements, there have been a number of observations which indicate that even abnormally coloured animals may remain together with the other members of their species. Meyer-Brenken, the head forester who studied the occurrence of melanistic roedeer *(Capreolus capreolus)* in the forestry department of Haste, discovered for example that there was no difference in the mutual behaviour of normally and abnormally coloured animals. When given a choice, both black and normally colour-ed males gave preference to females of the opposite colour. According to other observations the white females attracted greater attention from normally coloured roebuck than did other females. It has also been found that vixens prefer black foxes to normally coloured males.

Among the many obscurities that are related to the occur-rence of colour anomalies there remains first and foremost the question of their origins. Research is continually being carried out to find the links which might help to make clear the reasons why such colour deviations occur. For the time being it would seem that these links must be sought in several areas: in relation to the geographical distribution, the ecological conditions of the place of occurrence, dietary factors and finally physiological and genetic influences.

Even very strikingly coloured animals are sometimes tolerated by other members of their species. In this rare photograph, which was taken in the wild in northwestern Germany, the black and white spotted roe is grazing near a normally coloured roebuck. In the vicinity of these two animals a brown and white spotted fawn was also seen.

Probably the least important part is played by geographical distribution, even though it is known that in some places colour anomalies occur more frequently. For instance, albinism among carp is frequent in Japan, as is melanism among leopards *(Panthera pardus)* in South-East Asia. In Peru black specimens of the tamandua or lesser anteater *(Tamandua tetradactyla)* are extremely frequent and at the Vranov Dam on the River Dyje in Moravia several albino catfishes *(Silurus glanis)* have been either caught or observed. In some of these regions abnormally coloured animals predominate over normally coloured ones or even replace them completely. In this connection the frequent occurrence of melanistic roebuck in the plains of northwestern Germany is also noteworthy. The occurrence of colour anomalies in a large number of species is known in these regions, but it is not quite clear to what extent the actual geographical conditions affect their origin and to what extent local climatic, soil and other conditions are responsible.

These reflections regarding the influence of ecological conditions on colouring do not apply to colour anomalies only, but also to the origin of all coloration. It is generally assumed, for example, that a moist environment encourages dark coloration while a dry hot environment produces a tendency towards light yellowish red hues. In general most of the available evidence supports this view as is illustrated by desert animals which are usually of light coloration. This colouring applies not only to diurnal species for which the colouring might have protective significance but also to forms which are active at night and for which this kind of colouring would appear to be without value. Furthermore, many predominantly or completely black animals such as chimpanzees *(Pan troglodytes)*, black mangabeys *(Cercocebus aterrimus)* and black sakis *(Chiropotes satanas)* are inhabitants of moist forests. The best known such example is the black leopard *(Panthera pardus)* which is most frequently found in the equatorial rainforests of Malaysia and in the rainforests on the slopes of several African mountains. Similarly water-voles *(Arvicola terrestris)*, in which melanism is relatively common, live mostly in a moist environment. In the dry steppe regions of the Ukraine melanistic hamsters are almost unknown, while in moister areas there is a high percentage of black animals in the hamster populations.

A melanistic leopard *(Panthera pardus)*. Although at first glance the animal appears black, the oblique illumination highlights the underlying typical coloration of a black spotted pattern on a chestnut background.

In the dry hot sandy regions of Syria there lives a very light honey-coloured race of the brown bear *(Ursus arctos)*, the fur of which forms a striking contrast to the dark to blackish brown colour of Russian and central European races of the brown bear which are typically forest animals.

According to some zoologists, however, the moist warmth not only contributes to the origin of melanism, but in general to the production of various colour deviations. This is indicated among other things by the fact that those species of viper which occur in dry localites have constant coloration, whereas the adder *(Vipera berus)*, which lives especially in hilly and moist peaty localities, occurs not only in the black phase, but also in numerous other colour deviations.

Apart from these circumstances, which have not yet been completely explained, light conditions would also seem to have an influence on the origin of dark or light coloration. It is well known, for instance, that animals living in dark places, such as the cave-dwelling amphibian *Proteus anguineus*, are albino. But if the young of this species are kept in the light, their skins become pigmented.

Dark coloration, whether it be of the whole body or only a part (especially the head), can also provide protection against radiation, particularly that of ultra-violet rays. This is perhaps the reason why arctic animals also have black pigmentation of those areas which are most exposed, such as the snout or the eyes.

The influence of nourishment on animal coloration would seem to be confirmed, at least in some cases, by analysis of the diet. For example, melanism, or to be more exact a very dark colouring, can be produced in the bullfinch *(Pyrrhula pyrrhula)* through constant feeding with hempseed, and a darker colouring in European bison *(Bison bonasus)* can be achieved with fodder consisting of a large amount of oak bark. The coloration of red squirrels *(Sciurus vulgaris)* depends to some extent on the type of woods in which they live.

Biological, in particular physiological and genetic influences probably play a large part in the origin of colour deviations. All these anomalies apparently originate as mutations, that is, sudden but permanent hereditary changes caused by a modification of part of the genetic material. They are generally inherited as a recessive characteristic, that is, one expressing itself only in a homozygote constitution. (This is explained in the next chapter).

As previously stated, abnormal coloration does not necessarily occur in all siblings and it does not have to occur at all in the young of abnormally coloured parents. Normally coloured young have been born, for example, to albino females of the Canadian porcupine *(Erethizon dorsatum)*, to the koala *(Phascolarctos cinereus)* and to the albino roedeer. On the other hand, normally coloured parents can have abnormally coloured young. A female hedgehog *(Erinaceus europaeus)*, for example, once produced two albinos in a litter of three and on another occasion produced one albino in a litter of four. A pair of Himalayan bears kept in the Dresden Zoological Garden around 1890 regularly gave birth to albino young. One female hamster had three melanistic young in a litter of five, and two albinos were once found among twenty-one young in a hamster's nest.

It is not fully known to what extent these physiological and genetic influences are spontaneous and to what extent they are supported by climatic, dietary or other external factors. However, if one does not take this wider context into consideration it is extremely difficult to explain separately the various partial questions concerning animal coloration. This, by the way, is true not only of coloration, but also of most of the problems with which zoologists are faced.

COLORATION AND GENETICS

The genetics of animals is not a subject that can be explained and understood within the number of pages available in this book. However, if animal coloration is to be studied from as many points of view as possible, this approach must not be overlooked. The previous chapter touched upon several concepts used in genetics and these must now be more fully explained since without them there would be no basis for further development of the subject.

The material basis of heredity in all animals is to be found in the genes, which are carried on chromosomes. Genes determine the hereditary component of every feature of each individual, including coloration. All genes are paired and form what are called allelomorphic pairs. Before fertilization, that is before the creation of a new individual, the pairs of genes of the parents divide and the new individual receives one from each of the parents. This means that his hereditary constitution incorporates characteristics from both parents. Which of these characters is expressed in him depends on whether the factor concerned is a dominant or a recessive one.

A dominant character is one that is expressed even when it is present in a pair with a different character passed on by the other parent. On the other hand, a recessive character is only expressed when both constituents of the allelomorphic pair are the same. This can be demonstrated by means of the example of the heredity of colour and pattern in cattle, which

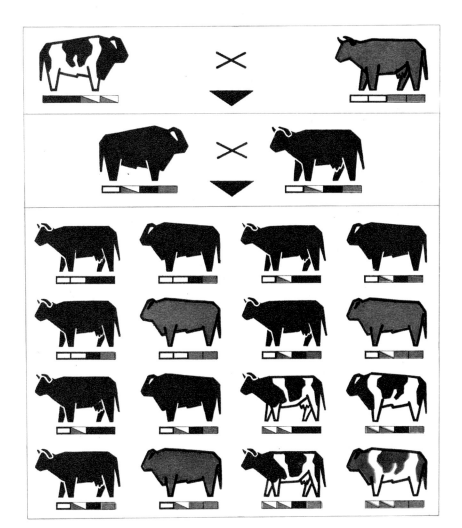

A diagram showing the heredity of two colour characteristics (colour and pattern) in domestic cattle.

From a black spotted bull and an unspotted brown cow (first row) the first generation of off-spring (second row) are all black. Through fur-ther crossing, the second generation of offspring comprises four types of coloration (third — sixth rows). The combination of character pairs is illustrated diagrammati-cally under each animal. A black oblong represents the factors for black, a brown oblong for brown, a white oblong for un-spottedness and a brown-white oblong for spot-tedness. The dominant factors (black and un-spotted) are indicated by a surrounding thick line. (Modified after Lan-precht.)

is illustrated in the diagram above. Of the pair of colour characters, black is dominant and brown recessive. In the pattern a single colour (unspotted) predominates over spot-tedness. The dominant character of any of the parents, that is black and unspotted, is always expressed in the coloration of the offspring, even when it forms a pair with a recessive charac-ter (brown and spotted) from the other parent. On the other hand, brown colouring and spottedness only appear in the coloration of the offspring when they are passed on from both parents.

This, of course, does not mean that both parents must be coloured in this way. From the diagram it is clear that in the first generation of offspring from a black spotted bull and a brown cow, all the animals are unspotted black. The dominant characters of both parents are expressed in them, (black and unspotted), but they do also have in their hereditary constitution the recessive characters of the parents, (brown and spotted) which are not evident from their external appearance. They are therefore not 'thoroughbreds' and their young will not all be black, but a combination of all the characters that occur in the hereditary constitution of this black pair. Four types of coloration will be found in the offspring: unspotted black, unspotted brown, spotted black and spotted brown. In comparison with the original parental pair, two new types of coloration appear here.

And yet only four of this second generation of offspring are 'thoroughbreds' (homozygotes). In the illustration these types are depicted on the diagonal running from the top left-hand corner (unspotted black) to the bottom right-hand corner (spotted brown). If each of these offspring is mated with a homozygous individual of the same genetic constitution so far as colour is concerned, all their offspring will be exactly the same colour and they will all breed true.

All the remaining twelve offspring of the second generation are heterozygous for at least one of the colour characteristics. This means that if any one of these animals is crossed with an individual of the same colour and bearing the same hereditary constitution, only some of their young will display the parental coloration. Of these, once again, only some will be homozygous and at least half will be heterozygous.

The above-mentioned example is one of the simplest, since there may be many more characters influencing colour. In some cases genes may interact in producing colour as in the process of epistasis. In this case the genes of one allelomorphic pair have the ability to suppress or block the effectiveness of one or more other allelomorphic pairs. According to whether this ability is possessed by a dominant allelomorph or by the recessive one, this is known as dominant or recessive epistasis.

Domestic pigeons living in a semiwild state in many cities resemble their wild ancestor the rock dove *(Columba livia)* in many ways. One characteristic feature of rock doves is the black wingbands.

The example of common pigeons can be used to demonstrate the action of dominant epistasis. Pigeons have, in all, five pairs of factors influencing colour: for complete black coloration, for complete black coloration with blue colouring of the tail feathers, for black patches on the back and wings and two black bands on the wingtips, for two black bands on the wingtips and a 'dirty' colouring on the wings and back, and for general greyish blue coloration with two black bands on the wingtips. These factors control the final colour in the sequence given above so that, for example, even if there is just one single gene for complete black colouring present in the genotype (hereditary constitution) of the pigeon, the bird will be all black, regardless of which other genes are present. If this gene is not present or if it is a recessive homozygote, another gene is allowed to come into action — for black colouring with a blue tail, followed by the gene for the patches on the back and the two bands on the wings and so on. A pigeon with greyish blue coloration with two black bands on the wingtips is only produced if the factors for all the alternative kinds of coloration are present only as recessive genes.

Recessive epistasis, on the other hand, occurs only if the allelomorphic pair which produces it is present, that is, only in the homozygous condition. If the allelomorphic pair is heterozygous, the dominant partner comes into action. An

The tarpan *(Equus ferus)*, one of the extinct species which has been 'brought back to life' by cross-breeding. This photograph was taken in the Polish National Park in Bialowieźa.

example of this is the overall coloration (not the pattern) of the proliferous swordtail fish *(Xiphophorus helleri)* which through mutation has produced a number of colour-forms. The basic coloration of the forms of this species which live in the wild is greyish green and is determined by two allelomorphic pairs, of which one is responsible for the production of the melanins and the other for the production of the enzyme which activates the pigments. If the first pair of allelomorphs is present in the recessive condition the production of melanin does not take place, resulting in so-called 'golden fish', having only a yellow pigment. If the other allelomorphic pair occurs in the recessive form the necessary enzyme is not produced. Albino forms, which are practically colourless and have red eyes can occur even though they may bear genes for melanin production for if no enzyme is produced no melanin can actually be synthesized.

If these albinos are crossed with golden domestic forms, there reappears in all offspring of the first generation the original greyish green coloration of the wild form of the species — a reversion to the naturally occurring type. In the second generation, nine of the fishes are coloured greyish green, three are golden and four are albinos. Of these albino fishes only one is genotypically pure. In the remaining three the albino coloration is caused by recessive epistasis of the allelomorphic pair responsible for the production of the activating enzymes which mask any effect of the genes responsible for melanin production.

The types of heredity described above are generally applicable to all animals. In many species there are also mechanisms dependent on the sex of the animal bearing the particular genes.

Sex controlled characters, where genes are present in both sexes but are expressed only in one, provide one example. Secondary sexual characteristics, among which coloration is a conspicuous feature, are the most important characters under sex control. Also other colour characteristics which are not obviously sexual are inherited in a similar way. For example, in some males of the aquarium fish 'peacock's eye' *(Lebistes reticulatus)* there appears a special 'zebrinus' pattern which is produced by a dominant gene. The same gene may also be present in females, but here it does not produce the pattern as the male hormone which controls its expression is absent.

The crossing of various species of mammal is quite commonly carried out in zoos. The photograph shows the male of a tigon — a cross between a tiger and a lion — in the Bois de Vincennes Zoo in Paris. The pattern is interesting as in many respects it is reminiscent of the coloration of lion cubs. (See illustration on page 80.)

A zebroid, a cross between a horse and a Grant's zebra, in the Nürnberg Zoological Gardens.

A variation of this principle is provided by the sex influenced characteristics in which the expression of the heterozygous allelomorphic pair in the external appearance of the animal only takes place if the sex hormone of one sex is present. Two types of the Ayrshire breed of cattle are kept of which one has bright reddish spots and the other has dark brown (mahogany) patches. The mahogany colouring is dominant and if the allelomorphic pair is homozygous, both sexes may have this coloration; if the 'mahogany' allelomorphic pair is heterozygous, the mahogany colouring appears only in the males, the females being coloured red. If both colour types are crossed, all the bull offspring in the first generation are mahogany and all the cows red, whereas in the second generation of offspring three of the bulls are mahogany colour and one is red, while among the cows three animals are red and one mahogany.

This simplified account may have created the impression that all these types of heredity are found only in domestic animals or beasts bred in captivity, where it is possible to select individuals of the required colour characteristics. Before demonstrating with some examples taken from nature that this is not so, it is necessary to define two terms which are very often used in genetics: phenotype and genotype.

The genotype is defined as the total of all hereditary capabilities of a given organism, and in the example already considered this is the total of hereditary factors for the production of coloration. Reference to the diagram on page 61 shows that in the second generation of offspring there are nine different types of character pairs, that is, nine genotypes. However, the dominant and recessive characteristics in these genotypes are combined in such a way that only four types of coloration are obvious externally. The four types of colouring which are evident by mere observation, are called phenotypes. The phenotype is thus the total of visible characteristics, regardless of the genetic constitution.

These technical terms are commonly used in breeding. In addition to the usual experiments with the crossing of various species of animal such as a lion and a tiger, a horse and a zebra, a polar and a brown bear and so on, many zoos have for a long time been carrying out experiments in order to reproduce extinct forms of animal life. Of course, it is not a question of actually renewing them, but of breeding animals which, in their appearance, are a true imitation of the extinct animals. So far successes have been achieved with two extinct ungulates — the aurochs *(Bos primigenius)* and the tarpan *(Equus ferus)*. Both these 'species' were bred by the repeated crossing of modern primitive domestic breeds. The resulting animals are not genetically speaking pure species but only phenotypes of the respective extinct forms. It is simply that in their external appearance they resemble species which are no longer living.

In the wild the inheritance of colour characteristics would not appear to be demonstrated in a very striking manner. The words 'would not appear' are used intentionally, as they reflect the actual situation very well. It has been illustrated, for instance, in the chapter on colour anomalies that they occur relatively frequently. At the same time they are mostly recessive mutants, so that they only become apparent if present

In very many groups of animals, for example the mammals, the coloration of the two sexes is usually the same. Many antelopes conform to this pattern although it is among them that one most frequently finds striking differences in the coloration of males and females. In the photograph is a male (left) and two females (right) of the algazel (*Oryx dammah*).

in a homozygous state. In addition to this, unlike plants or invertebrates, the tendency among vertebrates is towards comparatively few young so that usually not all the possible variations are found. The occurrence of colour characteristics in nature is thus actually very limited in comparison with the possibilities provided by a given genetic constitution.

However, the influence of heredity on animal coloration is evident in nature in many other cases, for instance in the crossing of closely related but differently coloured races and species. However, these are special cases which cannot be dealt with in detail here.

SEXUAL DIMORPHISM
IN COLORATION

Animal coloration is highly diversified and this variety is
not only manifested in the characteristic colouring of the
individual species, but also in differences within the species.
These are mainly sex differences which can be observed in
various individuals of the same kind and age, and seasonal
differences that are manifested in one and the same animal at
various periods of time.

The sex differences are on the whole well known. Here it
is a question of either permanent or temporary differences in
the coloration of males and females of the same species. The
immediate cause of temporary changes in coloration may vary.
Changes of colour are either caused by renewal of the body

The female (left) and
the male (right) of the
Carolina duck *(Aix
sponsa)*.

covering as in the growth of new feathers, or, in the non-repro-
duction period, the relevant colour pigments in the chromatophores may contract or be covered by other chromatophores.

It is assumed that the stimulus for these colour differences
is given by the activity of the sex hormones. This is borne out
not only by the occurrence of different coloration in the two
sexes out in the wild, but also by experimental work. In many
animals, for example, the differences only occur at breeding
time, when there is increased activity of the sex organs. The
red coloration of the belly of male three-spined sticklebacks
(Gasterosteus aculeatus) or of minnows *(Phoxinus phoxinus)*, the
bright coloration of some weaver birds (Ploceidae) and the
temporary black colouring of the beak of the male house
sparrow *(Passer domesticus)* are typical examples.

Basically the same effect has been achieved by experiments
involving the castration of animals or with the application of
sex hormones. For example, the darkening of the beak in house
sparrows can be produced artificially even outside the reproduction period by injecting the male sex hormone testosterone
and even females' beaks turn black after the injection of this
hormone. On the other hand, the typical yellow colouring of
the male blackbird's beak cannot be produced in a similar
manner in blackbird hens.

An interesting phenomenon occurs if a bird's sex organs
are removed. If a drake is castrated it permanently gains the
plumage of the female, and if a duck is castrated it acquires the
colouring of the male. The explanation of this strange effect
of castration on both sexes is provided partly by the physiology

The male (left) and the
female (right) of the
nyala antelope *(Trage-
laphus angasi)*. When there
are sex differences in the
coloration of antelopes,
the males are generally
dark brown to black and
the females light rust
brown.

Though the dwarf forest buffalo *(Syncerus caffer nanus)* is almost one single colour, the strikingly long hair of its ears has a conspicuous light pattern.

A diagrammatic illustration of the principles of heredity in budgerigars. The basic coloration of budgerigars is produced by two pigments, a yellow lipochrome deposited in the upper layer of the feathers and a dark melanin in the deeper layers. The action of these pigments is promoted by two allelomorphic genes, one pair initiating the production of the lipochrome, the other regulating the darkening of the melanin. If the first pair only are present in the dominant condition, the basic coloration is yellow. When both pairs are dominant, interference of light causes a green coloration. Interference operates in a similar way if only the second allelomorphic pair is dominant, and a blue colour is produced. If both allelomorphic pairs are present in the recessive state neither the lipochrome nor the melanin is produced and the result is a white coloration.

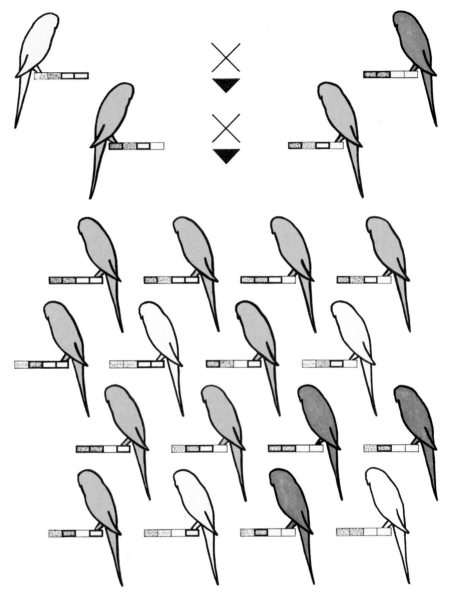

The top row of the drawing shows the parents. The second row shows the first generation of offspring and in rows three to six are the second generation of offspring. The yellow oblongs represents the gene for the production of the lipochrome, the dotted oblongs the gene for the darkening of the melanin. The dominant genes are indicated by a thick line around the relevant oblongs.

Ducks are a typical group of birds with striking differences in the coloration of the two sexes, especially in their breeding plumage. In the front of the picture are the female (left) and male (right) of the northern shoveler *(Anas clypeata)*. Behind them are the female (left) and male (centre) of the mandarin duck *(Aix galericulata)* and the male of the Baikal teal *(Anas formosa)* (right).

of the origin of sexual differences in coloration and partly by the anatomical peculiarities of birds' sex organs. If the proposition is accepted that the primitive coloration of birds is that of the female, whereas the colouring of males is produced secondarily by the effect of male sex hormones, it means that in the castrated male the female colouring occurs because the male hormones cease to have their effect. In the castration of females, however, the situation is more complex. Generally in adult birds only the left ovary functions, while its partner usually remains undeveloped. If this functioning ovary is removed from the duck, the hormones of the stunted gland begin to take effect and through their influence the duck apparently acquires the coloration of the drake.

On the other hand, castrated mammals always acquire the colouring of the female. After castration, black males of the nilgai antelope *(Boselaphus tragocamelus)* gain a rust brown colour similar to that of the females (in which practically no change of coloration occurs as a result of castration).

The fact that in some species differences in coloration occur only during the reproduction period can be accounted for if the non-breeding season is looked upon as a sort of natural castration of males caused by the fact that activity of

Sex differences in the coloration of the fish *Puntius nigrofasciatus*. Above is the female, below is the male.

the sex glands in most species of animal is very limited in the non-reproductive period.

Differences in coloration between males and females are not of widespread occurrence. Whereas in some groups of animals they are very striking, in others they are inconspicuous and in others again they are not emphasized at all.

There are many animal species in which the sex cannot be determined by the coloration. It would seem, however, that some groups of animals have a greater tendency to produce the same colour in both sexes, while in others differences occur relatively frequently. Mammals, and to a great extent reptiles and amphibians, belong to the former group. Great differences in the coloration of males and females are not found in mammals and when these differences do exist they are generally not very striking. They occur to a greater degree in some marsupials, monkeys and even-toed ungulates. For example, females of the red kangaroo *(Macropus rufus)* are usually grey and the males reddish brown. These differences, however, are not equally striking in all races and are almost unobservable in some. Adult males of the sacred baboon *(Papio hamadryas)* are silvery grey and the females greenish brown; the male of the black gibbon *(Hylobates concolor)* is black and the female light grey; the male black lemur *(Lemur macaco)* is also black, while the female is reddish brown.

The biggest sex differences among mammals are to be found in the colouring of even-toed ungulates such as the nilgai antelope, the Indian blackbuck *(Antilope cervicapra)*, the sitatunga *(Tragelaphus spekei)*, the bushbuck *(Tragelaphus scriptus)*, the nyala antelope *(Tragelaphus angasi)*, and the banteng *(Bos javanicus)*. In all these species the males are usually black or dark grey, while the colour of the female ranges through various shades of rust brown. In addition to changes in the background colour, there are often changes in the pattern. The rust brown female of the sitatunga has a dark stripe along its back while on the dark-coloured male this stripe is light in colour.

In reptiles too, sex differences in colouring are neither particularly common nor very striking. For instance, the male of the green lizard *(Lacerta viridis)* has a turquoise blue head, whereas in the female the colour of the head is greenish, similar to that of the rest of the body. A similar situation obtains with the tailed amphibians (Urodela). The male of the mountain water-newt *(Triturus alpestris)* is bluish black dorsally while the lower part of the body is orange with dark patches, whereas both the upper parts and flanks of the female are brownish.

On the contrary to the groups mentioned above, the sex differences in the coloration of birds and fishes are very frequent and also very striking. It has even occurred that the male and female have been described as two different species, as was the case with the Australian bird of paradise *(Astrapia mayeri)* until

Differences in the coloration of young (left) and adult (right) pencil fishes *(Nannostomus marginatus)*. The right-hand picture shows a female (left) and three males (right).

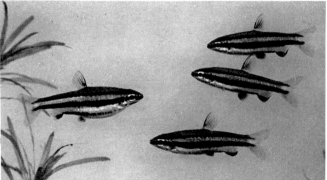

In some species ageing is revealed through the growth or enlarging of various bare patches on the skin. This diagram illustrates the differences of this kind in a young (left) and old (right) rook (*Corvus frugilegus*).

as recently as the nineteen forties. A similar case also occurred with tropical fishes of the genus *Cynolebias*.

However, the sexual differences are not equally striking in all species. Even the layman will notice some of them. He could distinguish at first sight the male and female of the chaffinch *(Fringilla coelebs)*, the brightly coloured drakes from the brownish mallard ducks *(Anas platyrhynchos)*, or pheasant cocks from hens *(Phasianus colchicus)*. In other species, often closely related to them, the differences are very slight. The partridge cock differs from the hen only in the horse-shoe-shaped patch on the breast and the colouring of the covering feathers in the shoulder joint. In budgerigars *(Melopsittacus undulatus)* the only sexual difference is the colouring of the cere, blue in males and brown in females.

The situation with fishes is similar. Fundamentally the same is true of them as of birds. The males are more motley and brightly coloured than the females and this can also be said of most other vertebrates in which there are differences of coloration according to sex. Only in exceptional cases are the females more strikingly coloured and it would seem that these reverse cases have come about mainly in those species where there has been a change in the usual manner of rearing the young.

Even though this external aspect of sexual differences in coloration is most interesting, it is not possible to describe it here in any detail and it should not be allowed to overshadow another much more important aspect of this whole group of colour differences, namely their function in the animals' lives.

AGE DIFFERENCES IN COLORATION

The term 'age differences' usually conjures up a picture of yellow goslings or the speckled young of seagulls but this is just one period in the life of animals, whereas age changes in coloration continue, even though in a much less striking form, throughout life. Not even animals are safe from ageing and in many of them, as with human beings, the grey-haired age is manifested in changes of coloration.

Just as with sex differences, age differences are not generally distributed among animals, but they are relatively more frequent and much more varied. It frequently happens that an animal progresses through several colour stages before reaching adulthood, some so different that the non-specialist considers each of them to be the colouring of a different species. In adulthood coloration changes much less and its overall character is not generally changed fundamentally by the process of ageing.

The coloration of young animals is very varied and its

Age differences in the roedeer *(Capreolus capreolus)*, as in some other mammals, are displayed in a lightening or a turning grey of the coat and an erosion of the sharp contrasts between the light and dark pattern. In the diagram the youngest animal is on the left, the oldest on the right.

origin is apparently influenced by many factors, from evolutionary and functional influences to relatively simple factors related to physical development. For instance, the young of fishes are often colourless when hatched and the pigment does not begin to appear in them until later. It usually appears first of all in the eyes, then on the yolk sac and its neighbourhood and finally spreads all over the body. This is true, for example, of herrings *(Clupea harengus)* or eels *(Anguilla anguilla)*. But the young that remain in the egg for a long time, for instance the genera *Fundulus* and *Aphyosemion*, are hatched already pigmented. In other fishes the young are pigmented, but in adulthood the pigmentation is lost. The American fish *Typhlogobius californiensis*, which when adult lives in holes dug out by crabs, is practically colourless, but its young, which live in free-flowing water, have a dark pigmentation.

In general it can be said that with age colours lose their lustre, bright colours grow darker and patterns lose their sharpness and distinctiveness. In many species, on the other hand, various features such as bald spots grow and are emphasized. In some mammals ageing is manifested in a very familiar way — the hair turns grey. This affects, for instance, domestic dogs in which the hairs on the head, especially in the region of

Young cuckoos *(Cuculus canorus)* have a conspicuous white patch at the back of their necks.

the mouth, turn grey with age. In old male gorillas the whole of the back of the body acquires a silvery-grey coloration.

Colour changes in the first period of an animal's life — from birth to adulthood — are more complex. Here it is only possible to compare the juvenile colours with the coloration of adult animals.

The simplest case is of course agreement or at least approximate agreement between the coloration of young and parents. This primarily is the case with various species of mammals in which it would seem that distinctive age differences in coloration are on the whole infrequent. However, the same coloration is commonly encountered in young and adults of other vertebrates, ranging from birds to fishes. Of course, it does not always have to be exact agreement. A lighter or darker shade in the coloration of young in comparison with that of adults is not a fundamental difference and neither are minor modifications of the pattern so long as they remain within the limits of the normal variability of the species. In some species where the adults of the two sexes are differently coloured, the young all resemble one of the parents, usually the mother, in colour. In these cases juveniles of one sex might appear to have the 'wrong' coloration. The fact that juveniles are more likely to take the maternal coloration is in accord

In the coloration of young lions *(Panthera leo)* there is a relatively dense spotted effect which is especially striking in some cubs.

with the now generally accepted view that the colouring of the female is the more primitive. The brighter coloration of the males, which is considered more advanced, is normally acquired by young males during the period of sexual maturation.

Apart from these minor variations some young have special colour characteristics which are lacking in the coloration of their parents, even though the general colouring of the young is the same as that of the parents. For instance, a young stoat *(Mustela erminea)* has a white rim round its ear lobes and a young cuckoo has a large white spot at the back of its neck, though its coloration taken as a whole is exactly the same as that of its parents. These colour characteristics may have a signalling function.

A second group of age differences in coloration includes cases where the young are fundamentally different from their parents in colouring, and look instead like adult members of other related species. The best-known examples are again to be found among the mammals and are shown by the stripes of young tapirs *(Tapirus indicus)* or wild boars *(Sus scrofa)* and the spots on fawns, young pumas *(Puma concolor)* or lion cubs *(Panthera leo)*.

In some cases this coloration is reminiscent of the pattern of other still extant but evolutionarily more primitive species of a particular group. For instance, spots like those on fawns are also found permanently on adult axis deer and Dybowski's deer. In other cases, for example in lion cubs or young pumas, the coloration bears witness to an earlier primitive type of pattern and thus indirectly supports the view that in these species monochrome colouring is a more advanced characteristic resulting from specialization. It is interesting to note that in some cases, particularly in lions, the spots last long into adulthood.

The third group of age differences in coloration consists of cases where the colouring of the young is not apparently related to that of either parent or of other related forms. At the same time, the coloration of the young may change

The change in coloration of both chicks and adult birds is caused by a change of feathers. The picture shows a young sparrowhawk (*Accipiter nisus*) moulting.

several times and each coloration may differ greatly from the previous one. Often the coloration may have a protective function, but sometimes it lacks any visible functional significance.

Such differences are encountered among birds and fishes. The speckled plumage of seagull chicks must undergo several changes before the young acquire the familiar coloration of their parents. Similarly the snow-white young of birds of prey generally pass through an additional juvenile stage before they achieve adult plumage. Young mute swans *(Cygnus olor)* are brownish, young crested grebes *(Podiceps cristatus)* are striped and the young of the water-rail *(Rallus aquaticus)* are quite black. The pattern on young fishes also generally differs radically from that of adults and their coloration is often brighter and more distinctive than in adulthood. For example, the fishes *Oplegnathus conwayi* and *Platax pinnatus* are almost the same dark colour all over when adult, but the young of the former are golden-yellow with two transverse bars and those

The young of some species of bird must moult several times before they acquire their adult plumage. This photograph taken on the Faroe Islands shows young gannets *(Sula bassana)* at various stages. An adult female is standing beside the youngest chick in the foreground.

of the latter are coloured yellow or orange, also with transverse bars. The flyingfish *Cypsilurus furcatus* is bluish grey, but its young are light yellow with black, yellow and orange-brown spots.

In other groups of vertebrates pronounced differences occur much less frequently. Among reptiles, for instance, such a difference can be found in the coloration of the green python (*Chondropython viridis*) from New Guinea, which is grass-green, while the young are coloured yellow or red. The young of the East African lizard *Mabuia quinquetaeniata* have bright blue tails which contrast with the colouring of the rest of their bodies, but in the adult lizards this colouring disappears. Among mammals the most striking age differences of this kind would seem to be the colouring of some seals whose young, unlike their dark-coloured parents, have white skins.

It is now necessary to return once again to the factors related to physical development which influence coloration, in order to show that such an apparently simple thing as

The coloration of the Kenyan giraffe (*Giraffa camelopardalis tippelskirchi*) has characteristic linked 'leaf shaped' patches (on the left). The right-hand photograph shows the reticulated giraffe (*Giraffa camelopardalis reticulata*) with its young.

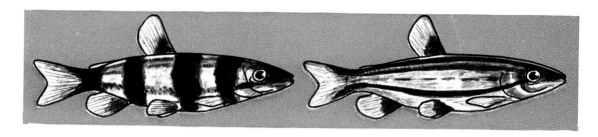

differences in the colouring of young and adults can be of great assistance in arriving at more broadly based conclusions.

Everyone is familiar with the coloration of the giraffe *(Giraffa camelopardalis)*, yet this coloration is not uniform, and is subject to quite distinctive changes within the various races. The reticulated giraffe, which inhabits the northern parts of the area of distribution of the species, has a coloration consisting of a dark brown background on which there is a light-coloured narrow network-pattern, while in the races living to the south these light interspaces are broader, the final result being the ill-defined fragmentary pattern of the South African Cape giraffe.

In spite of this, the coloration of young giraffes is basically similar in all races and is reminiscent of the pattern of the reticulated giraffe; thus the differences in coloration in the various races do not occur until adolescence. As the colouring of the young is considered in these cases to be more primitive, one can conclude that the original type of coloration in the giraffe is that of the reticulated giraffe, all the other types of pattern being of secondary origin.

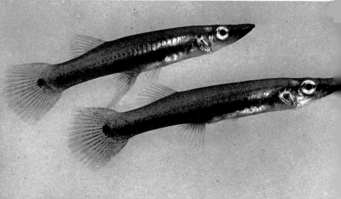

SEASONAL VARIATIONS

Whereas age differences in coloration usually occur only once in a lifetime, seasonal changes recur at more or less regular intervals. The term 'seasonal changes' suggests differences caused by renewal of coat or feathers brought about by climatic conditions, as in the case of the stoat *(Mustela erminea)* or wild mallard ducks *(Anas platyrhynchos)*. Seasonal changes of this kind are most characteristic of mammals and birds, but they are not restricted to these two groups and the manner in which they come about may also vary.

In a certain sense the coloration of reptiles, for example, also changes regularly, partly because shortly before the animal loses its skin the old one grows darker and partly because the new skin is generally coloured somewhat more brightly than the old one, even though it preserves the same pattern and colours.

Interesting periodic changes in coloration can also be observed in some fishes, not in connection with the alternation of the seasons, but as a result of the alternation of day and night. The fish *Latris hecateia* from Tasmania has olive green bars along its body during the day while in the night it has transverse dark bands. Similarly there are fishes from the genera *Nannostomus* and *Poecilobrycon* which are striped along their bodies during the day, but have transverse oblique bars in their coloration at night. Other fishes, in particular those which are active during the daytime, merely lighten their

The stoat *(Mustela erminea)* is a mammal which in the northern part of the British Isles shows distinct seasonal differences of coloration. In winter it has, apart from the black tip of the tail, a pure white coat.

coloration at night. In some cases these changes are related to the varying intensity of light, but in many species colour changes occur even when light and heat conditions are quite constant.

In a certain sense changes of colour which are related to the beginning of sexual activity are also considered as seasonal changes, for example in a large number of fishes such as salmon *(Salmo salar)*, bitterling *(Rhodeus amarus)*, minnows *(Phoxinus phoxinus)*, and sticklebacks *(Gasterosteus aculeatus)*, as well as in other groups such as ducks, where the males are brightly coloured in their breeding plumage but on the whole are similar to the females when in their eclipse coloration.

These 'sexual' seasonal changes also affect some mammals. The yellowish brown male of the squirrel *Sciurus finlaysoni* is coloured salmon red during the breeding season and yet the coat is not renewed at this period. The colour would thus seem to be the result of some secretion.

The great majority of seasonal changes, however, are linked at least to all appearances, with the climatic conditions in which the animals live. Consider the seasonal changes in

the coloration of snow hares, polar foxes and ptarmigans, all species which live in an environment with great differences between summer and winter. In general such differences are related to the protective function of coloration.

Several circumstances actually support such an explanation. For instance, the extent of the colour changes in members of the same or closely related species varies in relation to the climatic conditions in which the animals live. In the northern parts of its area of distribution, the arctic fox acquires an almost white coat in winter, but only rarely does so in Iceland. Similarly the weasel, which also becomes white in

Changes in the colour of the stoat's coat occur gradually during moulting, as can be seen from the diagram on the left. (Drawing modified after Psenner.) On the right is a photograph of the stoat in its brown and white summer coat.

winter in the north of Europe, only rarely has this colour in England and central Europe and when colour changes do occur they are generally limited to patches of varying sizes. Not even the stoat, which regularly turns white in central Europe, acquires a completely white winter colouring everywhere. In the south of the British Isles, for example, where there is a milder maritime climate, the white winter coat is uncommon and incomplete.

The relationship between seasonal changes and the habitat of animals is also borne out by another example from a geographically distant area. In Eastern Asia there live two closely related races of deer — the sika deer and Dybowski's deer. In summer members of both races have white spots, but in the winter period only the Dybowski's deer retain this spotted coloration. The sika deer, which live in leafy forests with deciduous trees, acquire a uniform dark brown colour at this time of the year.

There are, however, other equally closely related species in which seasonal coloration is manifested in different ways despite the fact that they live in very similar habitats. For instance, among lemmings there are species which turn white in winter, such as the collared lemming of North America *(Dicrostonyx torquatus)*, and closely related forms such as *Myodes lemmus*, where such changes do not occur at all. In the case of the rodent *Phodopus sungosus*, populations living in various regions behave in different ways and while one has the same colouring throughout the year, the other moults in winter to a practically pure white.

Experiments have been carried out on these same rodents to show what actually causes the change of coloration. According to the findings recorded, it appears that the moult into the winter coat is not so much influenced by change of temperature as by the relationship between the length of day and night, or to be more exact the relationship between the periods of light and dark.

There are, however, animals such as chamois *(Rupicapra rupicapra)* and some members of the deer family, whose coats grow dark in winter, and a great number of other species,

The coloration of beasts of prey is frequently influenced for transitional periods by the colour of the blood of their prey. In the photograph is a cheetah *(Acinonyx jubatus)*.

Young cheetahs differ from their parents in that they have a darker basic coloration of the coat and a mane of long light hairs growing on the neck and the back.

The European salamander *(Salamandra salamandra)*. The photograph shows the variability in the coloration of individual animals and at the same time the warning effect of the combination of yellow and black. The parotoidal glands which produce a poisonous secretion can be clearly seen on the sides, behind the head.

especially birds, where there are seasonal changes of coloration having no apparent connection with the ecological conditions of their habitat.

Thus seasonal changes in coloration do not have any common cause and in spite of external similarities determined by the regular alternation of the seasons, they are very diverse in their nature. One must include here differences of colour caused by a renewal of the skin, sexual differences that are manifested mainly in the social life of animals, and differences dependent on the varying ecological conditions in which individual species live. Finally one must include changes caused by the physical wearing out of the body covering such as the snapping off of feather tips. Thus it has been demonstrated that it is not sufficient merely to describe and compare the various types of animal coloration. In order to grasp their significance one must learn about their function and the way coloration is related to other aspects of animal life.

PROTECTIVE COLORATION

It would be difficult to find a subject better-suited to the beginning of the second part of this book than the question of protective coloration in animals. The concept of camouflage or cryptic coloration is one which is firmly ingrained in the minds of most people who remember from their school days examples of animals which melt into their surroundings. But while this idea has become fixed in the public consciousness, the opinions of experts on this matter are not completely undivided. There are many who maintain that coloration has no protective function at all and who consider any explanation of this kind to be typically anthropomorphic, caused as it is by an exaggerated emphasis on our human view of the role of coloration and on our general manner of orientation within our environment.

First of all it is necessary to elucidate the term 'protective coloration' itself and the principles on which it is based. Protective coloration means that type of colouring which provides the animal with a certain protection and which helps it either permanently or in some situations associated with its way of life. The help provided by protective coloration, however, may be of various kinds.

Camouflage, as the name itself indicates, must conceal the animal or make it inconspicuous. Most people think of the camouflage effect of coloration being achieved by the motionless animal merging with its surroundings. Most animals, how-

The mountain hare *(Lepus timidus)* on a snow covered plain in Greenland.

ever, are very active and so this passive, though effective protection would not be of much help. If camouflage is to be effective it must conceal the animal both at rest and when in motion, for such coloration is possessed not only by hunted animals but also by beasts of prey which need to approach their quarry as inconspicuously as possible. It is thus not only a matter of the animal remaining unobserved, but also of the animal being difficult to identify should it have to move. A tennis ball in flight can be seen easily because it is white, but if it were coloured red, it would be much more difficult to follow. This principle is also applied in animal coloration. The forms that are to be concealed are coloured, as far as is possible, in the same way as their surroundings while those which have to draw attention to themselves are strikingly coloured with contrasting colours that are most unlikely to occur in the surrounding environment. These species generally have effective means of protection like poison glands, evil smelling secretions, or strong claws and their coloration is basically a warning to the prospective opponent of an unpleasant encounter.

Another principle which is found relatively frequently in animal coloration is the direction of attention away from vitally important organs or parts of the body to less important parts. In many species of lizard, for example, the tail is coloured more strikingly than the rest of the body. The lizard draws the enemy's attention away to that part of the body which, if lost, does not constitute a threat to its life. The East African lizard *Holaspis guentheri* has a beautiful bright blue tail that contrasts sharply with the black and bronze-coloured stripes on the body. Sometimes this coloration lasts throughout the animal's lifetime. In other cases it is most conspicuous in the young which are probably not as good at escaping from their enemies as are adult animals.

In many preying species, on the other hand, the pattern draws attention away from the most dangerous part of the body and serves as a bait for the quarry. Some young rattlesnakes *(Ancistrodon contortrix, Ancistrodon piscivorus* and *Bothrops atrox)*, for example, have bright yellow tips on their tails. Thus they draw attention away from the head and at the same time attract the attention of their prey with the bright colour. Many animals use excrescences in the region of the mouth to attract their prey. The Mediterranean fish *Zaleoscopus tosae* sometimes makes use of a protuberance on its lower jaw. Its shape resembles a little red thread which wriggles in the sand and imitates the movement of a small worm. Some turtles have similar coloured protuberances on their tongues — for instance,

Fishes, such as *Pterophyllum eimeckeri*, shown in the photograph on the left, often possess a pattern of transverse bars, the function of which is explained by the diagram (bottom right). The first drawing illustrates the basic appearance of the fish, the following two the effectiveness of disruptive coloration against single-coloured backgrounds and the last one demonstrates the effectiveness of disruptive coloration in the natural habitat. (Drawing after Cott.)

Chelus fimbriatus and *Macroclemys temmincki* lie in the water with their mouths open and attract their prey with white excrescences.

An interesting combination of the two principles mentioned above is a pattern that disorientates the enemy, camouflaging important organs and creating the impression of other organs at the opposite end of the body. For instance, many animals have an 'eye' pattern in a completely different position from that of the real visual organ. The real eyes are then camouflaged as in many fishes and frogs.

These are essentially the principles on which the protective function of coloration is based. The actual manner in which the protective effect of coloration is achieved, however, varies enormously. Because the various principles which are applied in animal coloration are so interesting, it will be well worth while pursuing this subject and describing some of the elements of protective coloration in more detail.

Variations of the 'eye' pattern recur in many species of fish. The photographs show *Astronotus ocellatus* (left) and *Crenicichla lepidota* (right).

CAMOUFLAGE

As already explained in the previous chapter, the main purpose of camouflage is to conceal its bearer and make him inconspicuous. This effect is achieved in nature in many ways, sometimes even with means that seem contradictory.

The simplest case of camouflage coloration is general agreement between the coloration of the animal and that of the surrounding countryside. Some of the relationships between

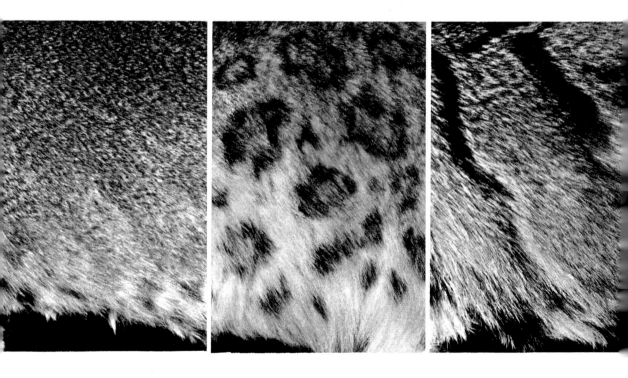

animal coloration and the creature's habitat have already been discussed so it is now necessary to find the extent to which agreement between the animal's coloration and the colouring of its habitat may provide protection in the sense described·in the previous chapter.

It is indisputable that correspondences do exist between animal coloration and the colour of the surrounding terrain. In some cases the close similarities are very striking, because they occur in forms that are not only distant from one another geographically, but are also only remotely related.

The basic coloration of animals living in desert or semi-desert areas, for instance, is very often a sand yellow, regardless of whether they are mammals such as the fennec *(Fennecus zerda)* or the Egyptian jerboa *(Jaculus jaculus)*, birds such as the sandgrouse (Pteroclidae) or bustards (Otidae), or reptiles such as the desert monitor lizard *(Varanus griseus)* or some agamas (Agamidae). Even more striking is the agreement in coloration of tree-dwelling creatures which usually have the same green coloration regardless of which geographical area they live in or to which class of vertebrates they belong.

The very closeness of the relationship between the col-

Close-ups of the coats of several species of preying cat show the range of colour adaptation to various types of environment. From left to right: the golden cat *(Profelis aurata)*, the snow leopard *(Uncia uncia)*, the clouded leopard *(Neofelis nebulosa)*, the lynx *(Lynx lynx)*, the jaguar *(Panthera onca)* and the tiger *(Panthera tigris)*. Each photograph shows a section of the coloration from the dorsal area down to the belly, from approximately the centre part of the body.

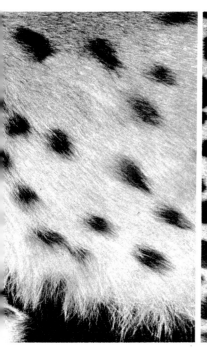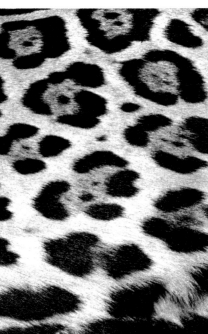

oration of some animals and the colour of the ground on which they live is noteworthy. This similarity is so striking in some forms that it occurs even in relatively small geographical areas. American iguanas of the genus *Phrynosoma* (often known as 'horned lizards'), for instance, are known for the differences that occur even in the coloration of individual races, differences which are related to the kind of soil upon which the animals live. *Phrynosoma douglassii ornatissimum* has a pattern adapted to the coloured cliffs of the Arizona Desert; the single-coloured *Phrynosoma douglassii douglassii* lives on soil of one single colour; the almost white *Phrynosoma platyrhinos* lives on white salt flats; *Phrynosoma blainvilli frontale*, which alone lives in a forest region, has the colour of fallen pine needles.

Similar local races are known among mammals too. Thus local forms of rodents from New Mexico differ from one another in their coloration in spite of the fact that they do not live far apart and are subject to the same climatic conditions. Whereas the desert forms have a light rust and reddish coloration, in close vicinity to them on a broad lava band there live the almost black forms of the pack rat *(Neotoma albigula melas)* and the pocket mouse *(Perognathus intermedius ater)* and as a close neighbour on a white sandy territory there lives the almost white *Perognathus gypsi*. The south-western region of North America is the home of many other coloured rodents with

This photograph of zebras at a watering-place shows how the striking and apparently very conspicuous patterning of zebras in reality makes them more diffcult to see. It is also difficult to pick out individual animals.

light, medium or dark coloration, living as neighbours on the corresponding light, medium or dark-coloured ground.

Such differences in the coloration of closely related forms, linked with life in various habitats, are called adaptive radiation. It is encountered in many groups of animals, for instance cats, to mention just one group of better-known animals. The coat pattern of some of them is shown in the photographs on pages 96 and 97. However, many readers will no doubt be struck by the thought, when looking at these photographs, that the pattern of the coat is too conspicuous to conceal the animal in the wild. But this is a false impression caused by looking at the coloration in isolation from the environment in which the animal lives. After all, zebras have just as striking a pattern as tigers and yet they are less conspicuous in the wild than many other animals. This is borne out by the observations of experienced hunters and travellers who have seen them in the wild. The traveller Stewart White, for instance, who 'saw thousands and thousands of zebras in all sorts of different surroundings' writes: 'For one reason or another, in the given surroundings the zebra is always less conspicuous than any other animal. The black and white stripes merge with the surrounding foliage so that the zebra cannot be seen even at a very short distance. Many times ... we couldn't even make them out at a distance of forty to fifty metres. Even our servant ... who was one of the sharpest eyed of the natives ... was deceived. And yet the countryside was so open to view that antelopes or gnus could be seen at a distance of 200 metres.'

The optical effect of striping depends to a large extent on the direction of the stripes. If the stripes run parallel to the outline of the body they emphasize its shape. On the other hand, stripes which run across the body outline help to break up the shape of the animal and enable it to merge into its surroundings. (Modified after Cott.)

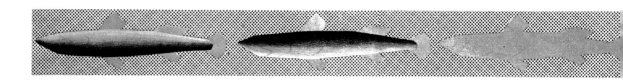

How is it then that such apparently striking coloration — stripes or patches — can conceal such large animals as zebras or giraffes? In order to answer this question, it is necessary to try to elucidate the essence or the camouflage effect in the various types of coloration.

As already indicated the simplest way of protecting an animal is for it to have the same colour as its habitat. This agreement would in itself be insufficient to conceal the animal. Even a single-coloured animal standing in front of an identical ground can be seen very well if it is illuminated from above as it would be in natural circumstances; whereas the back is over-illuminated, the underside is in shadow so that once again it creates the impression of a three-dimensional body. This impression of depth, which reveals animals more than their colour, is destroyed by so-called counter-shading. The following examples should help to demonstrate the nature and function of counter-shading.

It is generally known that animals are usually darker on their backs than on the underside of their bodies. The significance of this arrangement is explained by the diagram on this page. Since light mostly falls on the animals from above and this area is then over-illuminated, it is coloured darker so that the overall hue roughly corresponds with the lighter colouring of the underside which is in shadow and is therefore apparently darker. The result is an optical 'flattening' of the animal and the impression of a three-dimensional body is lost.

This explanation would seem to be highly speculative and far-fetched but its correctness is proved by a number of observations from nature and primarily by the peculiarities in coloration of those species which live in exceptional conditions. For example, most fishes have a lighter colour on the underside of their bodies but flatfishes (Pleuronectidae), in which the proportions of the body have undergone an interesting change and which are permanently turned over on their sides, are a lighter colour on the sides on which they lie. If the bottom of the tank is permanently illuminated, pigment is produced on the blind side of each flatfish's body too.

An even more interesting confirmation of the correctness of this way of explaining the function of counter-shadow is found in the Nile fish *Synodontis batensoda* which is coloured in the opposite way to most fishes — light on the back, dark on the underside. This apparent contradiction can be explained by observations in the field which have shown that this fish swims with its belly turned upwards — thus earning its common name, the upside down catfish. The reverse coloration therefore also acts as counter-shadow.

The greatest and most striking application of counter-shadow is found in animals exposed to great differences in the illumination of the back and the underside, in the surface strata of rivers and of the sea. Fishes living near the surface show

The coloration of the gaboon viper *(Bitis gabonica)* shows the principles on which 'dismembering' patterns are based.

If a surface is to be divided optically into several apparently separate areas, the most effective pattern is that which is based on the principle of the greatest contrast. Light and dark areas meet at a clearly defined boundary line and their intensity increases as they approach this boundary. (Modified after Cott.) The pattern of many snakes is based on the principle of the greatest contrast. The photograph on the right gives a close-up of the snake *Bitis lachesis*.

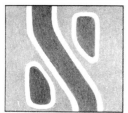

a striking difference between the coloration of the back and that of the belly, as do animals which live in the open country-side against a dark background, such as antelopes and rodents. On the other hand, there is very little difference in colour intensity between the backs and the undersides of desert animals since the undersides are illuminated by light reflected from the sand. Fishes that live in deep water or in muddy water giving a poorly illuminated environment without any strong contrasts also show little difference in colour intensity.

The creation of a counter-shadow, however, is not restricted to single-coloured animals. Any pattern of stripes or spots can achieve a similar effect by the pattern on the upper side being larger and denser and growing smaller and less well defined towards the underside of the body.

But the pattern has yet another function in camouflage coloration. It divides up the surface of the body optically, thus destroying the impression of a solid mass. The observer concentrates on the spots and the overall shape of the animal escapes him. Basically the effect is the same as that of net curtains in a window — the observer's gaze stops at the curtain and does not penetrate into the darker room. Though it is not impossible to recognize the animal in this case, one requires more time in order to distinguish it. The advantage of this type of pattern is that the animal does not have to keep to one particular colour environment and its radius of activity is larger than it would otherwise be. Of course, it is a good thing if part of the pattern is the same colour as the ground, but this is not absolutely necessary. More important is the main 'dismembering' pattern, the colour of which should contrast as much as possible with the predominant coloration of the ground. Dismembering coloration is more effective when the most strongly contrasting elements are positioned next to one another, so creating a sudden transition. Such coloration is relatively common in snakes, not only among the non-poisonous species such as the boas and the pythons (Boidae) but also in many less agile poisonous snakes such as the gaboon viper *(Bitis gabonica)*. However, the principle of strong contrasts is also applied in the coloration of other vertebrates. Among mammals the anteater *(Myrmecophaga jubata)* and the viscacha *(Lagostomus trichodactylus)* have such a pattern. Many amphibians, fishes and above all numerous groups of young birds also

Dismembering coloration enables young black-headed gulls *(Larus ridibundus)* to hide safely in reeds by the water's edge.

A diagrammatic illustration of the East African frog *(Megalixalus fornasinii)* in the normal position (left) and resting, when it folds its legs up close to its body (centre). The picture on the right shows the pattern of stripes drawn in only lightly, so as to indicate the way in which they run across the various parts of the body linking them up in an optical unit. (After Cott.)

have colour patterns which are among the most perfect examples of the various types of camouflage known in nature.

It has been observed that the pattern on the body of some animals has, in certain circumstances, another very interesting effect, in that it appears to link up various parts of the body. The diagram on this page illustrates the coloration of the East African frog *Megalixalus fornasinii* which has rust brown stripes along its back, flanks and legs, while the rest of the coloration is a silvery white. When this frog is sitting at rest, it has its forelegs tucked under its chin, its hind legs folded up close to its body and its eyes covered by the lower lids. In this position the coloured stripes of the various parts of the body link up in such a way that they seem to run uninterruptedly right along the body and limbs. The frog thus creates the impression of an indeterminate object, in which it is very difficult to recognize the body of an animal.

A similar linking pattern, though not so striking in form, is to be found on the hind legs of the common frog *(Rana temporaria)* and some other species of frog. In one of these species, *Edalorhina buckleyi*, the pattern on the hind limbs is most remarkable. The linking pattern is here formed by three dark stripes, two of which are narrow and the third broad. Their mutual distance from one another is the same on the thigh, the lower leg and the foot, but the sequence of the stripes on the lower leg is the reverse of that on the thigh and on the foot. What is the significance of this pattern? If the frog draws its hind leg up to its body, the various parts fold up in such a way that the lower leg is against the thigh and the foot is turned through 180°. If the arrangement of the broad and narrow stripes were the same on all three parts, they would not link up when the leg was folded, so although the stripes seem unrelated when the legs are extended, they fit together when the limbs are folded.

A similar linking pattern is also encountered in other groups of vertebrates. In many snakes such as the East African viper *(Vipera superciliaris)* and the gaboon viper *(Bitis gabonica)* the pattern links the upper and lower lips. In fishes it often links the body with the fins and in all vertebrates this pattern makes it possible to a large extent to camouflage the eye which, because of its regular and sharply delimited round shape, is one of the most conspicuous forms in nature.

It is interesting to note how often and in how many different ways the pattern of the head either completely conceals the eye or makes it inconspicuous. In some species the whole eye is covered with dark patches of various shapes as in the red-backed shrike *(Lanius collurio)*, the badger *(Meles meles)* and the gemsbok *(Oryx gazella)*. In many species, particularly fishes, the dark patch covers only part of the eye. It generally covers the whole pupil and part of the iris, while the rest of the eye is adapted in colour to the colouring of the head. In both cases the deep-coloured patch draws attention away from the

The frog *Rana temporaria* The transverse brown stripes optically link the various parts of the folded rear limbs, and the large brown patches on the head camouflage the position of the eye.

borders of the eye and head and often redirects it to some other, less important part of the body. These patches vary in shape. They may consist of long vertical stripes as in the fish *Eques lanceolatus* or stripes that stretch right along the body, as in the fish *Trichogaster leeri*. Sometimes they are merely a system of lines which incorporate the eye as in the lion-fish *(Pterois*

Sex differences in the coloration of the Indian blackbuck *(Antilope cervicapra)*. This type of sex difference (the female lighter, the male darker) is found in many species of even-toed ungulates.

In the sable antelope *(Hippotragus niger)*, as in a number of other antelopes, the coloration produces a striking face mask.

A wide range of colours and patterns occurs in vertebrates yet this coloration is not always necessarily conspicuous when the animal is seen in its natural habitat. In the picture is the sea iguana (*Amblyrhynchus cristatus*) from the Galapagos Islands.

The greatest variety of coloration occurs in fishes both as regards pattern and the brightness and combination of colour. The photograph shows the male of the fish *Aphyosemion nigerianum*.

volitans) or largish irregular patches in which the eye is positioned asymmetrically as in the common frog *(Rana temporaria)*. Sometimes the eye is not covered at all, but in its vicinity there are a large number of small irregular dark spots which distract attention and thus conceal its position. This is apparent in the young of many birds, and also in some adult birds such as the great spotted woodpecker *(Dendrocopus major)*, and the woodcock *(Scolopax rusticola)* as well as in many snakes. The most complex camouflage is generally to be found in species without eyelids or in species where the lids are joined together such as fishes and snakes, as to them such camouflage is particularly important. In other animals it is the movable eyelids which very often serve to camouflage the eyes. Many birds such as nightjars (Caprimulgidae) have lids covered with feathers, while some reptiles such as chameleons have them covered with scales. In this group the borderline between the eye and the covering of the rest of the head practically disappears when the eyes are closed, especially as in some species like the spiny

Dismembering coloration in the frog *Ceratophrys appendiculata.* The impression is created of a flat object throwing a shadow on uneven ground.

The lionfish *(Pterois voli-tans)* showing camouflage of the eye in a dark stripe which runs through the eye. Only the pupil is hidden in this way, the rest of the eye conforming to the background color-ation of the head. The bright and apparently conspicuous pattern func-tions as dismembering coloration in the environ-ment of coral reefs. The photograph was taken at the Zoological Gardens in Berne.

lizard *Moloch horridus,* where there is a linking pattern of patches which run fluently from the upper lids to the lower ones when the eyes are closed.

Just as striking as the outline of the eye is the outline of the animal's whole body. Camouflaging it is more difficult as no pattern can be produced that would link all of it with its surroundings. One alternative for camouflage is an optical breaking-down of a coherent form, as is described at the begin-ning of this chapter. There are, however, more interesting ways of camouflaging the whole body. One of them is found in frogs, where the creation of an artificial shadow appears to flatten the body or helps to conceal it in grass or reeds. In the coloration of the sides and back of the quail *(Coturnix coturnix),* for instance, there are yellowish white spots over black inter-secting stripes that together look like shadows thrown by dry grass.

The outline of an animal may also be camouflaged in other ways, perhaps with an irregular fold of skin, but in this case the most important factor is not coloration, so it will not be dealt with in any greater detail.

WARNING COLOURS

While the aim of camouflage coloration is to make the animal inconspicuous, the purpose of warning coloration is the exact opposite. This kind of coloration points to the presence of the animal and as the effect is often emphasized by special distinctive behaviour it can make the creature easier to recognize. Most species coloured in this way have other effective means of protection which make them unsatisfactory prey.

A number of beasts of prey of the stoat family have a striking combination of black and white in their coloration and are well known for the fact that when in danger they squirt a foul-smelling liquid at their enemy.

Another effective weapon of animals with warning color-

The African honey badger *(Mellivora capensis)* illustrates warning coloration. The light dorsal surface and the dark undersurface emphasize the three-dimensional appearance of this animal, an effect exactly opposite to camouflage countershading. (Photograph from the Zoological Gardens in Basle.)

Warning coloration is sometimes only present on the underparts of the body, as for instance in the frog *Bombina bombina*, and is only displayed when the animal assumes a certain posture.

ation is poison. Either they emit a poisonous secretion from the skin glands on the surface of the body, or they have poison glands in the mouth and inject poison into wounds caused by biting.

Many amphibians belong to the first group. For instance, the secretion from the skin glands of the European salamander *(Salamandra salamandra)*, which has a typical bright yellow and black coloration, is particularly noxious. One of the most poisonous frogs anywhere is the South American *Dendrobates tinctorius* and Columbian Indians use the highly poisonous secretions from its skin to produce poisoned arrowtips. The coloration of these frogs consists of large white, yellow, red or light blue patches on a chestnut or black ground. One of the most brightly coloured species of African frog is *Phrynomantis bifasciata*, a dark grey or blackish frog with two orange or pink stripes on the sides of the body and patches of the same colour on the back and on the back legs. This species also emits a large amount of poisonous secretion when defending itself.

Perhaps the best example of the relationship between the coloration, the means of protection and the behaviour of animals, is shown by poisonous lizards. There are very few of them, just two species, the gila monster *(Heloderma suspectum)* and the Mexican bearded lizard *(Heloderma horridum)*, which inhabit the desert regions of America. Whereas most desert lizards are fairly agile, vigilant animals with camouflage coloration, both these poisonous species are slow and ungainly in their movements and have a conspicuous warning coloration consisting of a combination of pink spots on a dark ground.

Of course, one must understand what conspicuous coloration really is, for often an apparently conspicuous colour can actually conceal the animal very well in its natural habitat. Coral reef fish, for instance, are among the most brightly coloured animals, but it is a case of camouflage coloration which conceals them in the multicoloured submarine world of corals, sea anemones and water plants. The black and white colouring of the Abyssinian colobus monkey *(Colobus abyssinicus)* is also very striking in the museum show case, but in its natural environment it is almost completely lost among the black trunks of the trees and the long grey curtains of lichens on the branches.

Warning coloration is thus coloration which is extremely conspicuous in the animal's natural habitat. The patterns are usually simple and common to a number of species with warning coloration, so making for easier recognition. The position of the colours may be reversed with light shades dorsally and darker ones below and the colours may be frequently repeated in the same combinations of red-black-white, yellow-black-white, red-black, orange-black or black-white. The combination of

The African toad *Bufo superciliaris* is coloured in such a way as to imitate a shadow thrown by a flat leaf.

113

The shape and coloration of the South American leaf-fish *(Monocirrhus polyacanthus)* reminds one more of a floating leaf than of a living animal.

black and white is especially important for nocturnal animals because it provides the greatest colour contrast in twilight.

Not only the pattern, but also the colour combination may recur in a number of unrelated forms from various genera, families and even classes. For instance, the combination of yellow and black occurs in several species of salamander, the yellow-bellied toad *(Bombina variegata)*, the tree-snake *Dipsadomorphus dendrophilus*, the yellow-bellied sea-snake *(Pelamis platurus)* and many species of fish.

There are, however, many species in which the warning coloration is only displayed temporarily, usually when the animal is threatened. When at rest these species are often camouflaged and the strikingly coloured parts are easily concealed in the folds of the skin, covered with hair or feathers, or positioned in such a way that they only come into effect when the animal assumes a special posture. For instance, the yellow-bellied toad has warning coloration on the underside of its body. This coloration only becomes effective when it takes up its warning position by standing stock-still with its legs erect, or by turning over onto its back.

Many other species take advantage of this element too. Some snakes such as the black-necked cobra *(Naja nigricollis)* stand erect before the enemy and show the strikingly coloured horny matter on the undersides of their bodies. Similar behaviour on the part of the hamster *(Cricetus cricetus)* and the marbled polecat *(Vormela peregusna)*, which have pitch-black undersides to their bodies, fulfils the same function.

In some species the warning coloration is completely hidden when at rest. For example, the Siamese toad *Callula pulchra* assumes a warning position by blowing up its body, while at the same time on its back two broad folds of skin coloured bright yellow open up. A similar principle provides the basis for an interesting form of threat used by the African crested rat *(Lophiomys ibeanus)*. It is not a large animal and its coloration at rest is a very light grey. However, when it is angered, the hair on its back stands up and at the same time the hair on its flanks drops, so a number of stripes running from the head to the tail suddenly appear in its coloration. The change is produced partly by the two colours of the hairs, which are white at the tip but black at the base and partly by the dark grey colouring of the skin which is laid bare at the places where the erect and flattened hairs meet.

When faced with an enemy some salt-water fishes, such as gurnards, open up their brightly-coloured fan-like breast fins and produce a pattern. If the rays of the fins are folded, the pattern is hidden. A similar system is also employed by some birds such as owls, grouse and parrots, which achieve the same effect by spreading out their feathers.

The element of surprise can also be brought into action by the coloration of the mucous membrane which becomes conspicuous when the lips are parted or the whole mouth is opened. This form of warning is frequent in reptiles and exists, though less commonly, in birds. The colour of the mucous membrane is generally black, red, pink, orange or yellow. The East African agama *(Agama atricollis)*, which is protected by camouflage coloration while at rest, will put up and open its jaws when it is antagonized so that the orange-yellow colour of the mucous membrane suddenly becomes visible. A similar reaction is known in many species of snake.

DISGUISE AND MIMICRY

Imitation differs from other types of protective coloration to a certain degree, mainly because it does not introduce any new elements of colour or pattern, but emulates what has already been created elsewhere. An animal may either imitate part of the natural habitat, or it may copy the coloration of other species.

Passive similarity with some harmless object in which the enemy is not interested such as a leaf or a plant is relatively frequent. In the Lower Amazon Basin, there lives a smallish leaf-fish *(Monocirrhus polyacanthus)* which in its appearance and coloration is almost identical with a dead leaf. It is flat, has the shape of a leaf and also has a fleshy outgrowth below the jaws which resembles a leaf-stalk. Its coloration is very changeable, the basic colour varying between light grey, golden cinnamon and brown, each colour being overlaid with darkish, irregularly spaced spots. The eyes are camouflaged and on the lateral surfaces of the body darkish spots form an imitation of the central vein of a leaf.

This fish frequently lives in smaller rivers and tributaries with clear, shaded pools and in flooded forests, as it prefers still water. Here it either hangs with its head down or lies near the surface of the water with its head inclined down at an angle. It moves with the aid of colourless dorsal and anal fins, or allows itself to be carried along by the current. Sometimes it remains in the vicinity of the stems of water plants or algae

The leaf-tailed gecko (*Uroplates fimbriatus*) imitates the bark of a tree overgrown with lichen.

so that it appears to be caught in them; at other times it lies motionless on its side on the river bed where it is lost among the water plants and algae. J. B. Cott writes in his well-known book on protective coloration that in order to catch this fish one has to net many waterlogged leaves and study them carefully before finding what one is looking for. Even when the fish is caught, it lies flat and motionless like a dead leaf.

In fishes, however, such imitation is not on the whole too difficult, since in many species the body is already relatively flat. It is much more noteworthy that a similar adaptation — the imitation of a leaf — is found in other vertebrates where this flattening is practically impossible, as in frogs and reptiles which have a relatively voluminous body. In these animals the impression of flatness is created for the most part by the pattern. For instance, the South African toad *Bufo superciliaris*, which is flattened on the dorsal side of the head and body, has folds of skin which run from projections above the eyes along the sides of the body to the rear end, more or less dividing the body into upper and lower parts. On top the toad is brown or grey with darkish spots, but under the folds of skin the body has a clearly defined dark chestnut colour. This creates the impression of a sharply delimited shadow thrown by the edge of a leaf.

Animals from many other groups imitate the bark of

The long-headed snake (*Oxybelis acuminatus*) can easily be mistaken for a liana at first sight.

trees. Such coloration is found amongst such diverse groups as treefrogs, iguanas, geckoes, wry-necks and nightjars as well as in many other species. In North America, for instance, there lives a small treefrog *Hyla femoralis* which in its colour and pattern exactly imitates the bark of the pine tree. It is also highly dependent on these trees and does not occur in places where pine trees are completely absent.

Aquatic creatures, primarily fishes, also imitate seaweed, and pelagic species such as sargassumfishes *(Histrio histrio)* provide good examples. Here of course the form of the body plays a greater role than the actual coloration. Typically these forms are yellow with irregular dark bands on which are small white spots representing the organisms living on the surface of the seaweed.

Imitation can run to incredible extremes. Many snakes, for instance, imitate lianas both in their coloration and in the shape and breadth of their bodies. In the forests of the Amazon there lives a green, brown or greyish coloured snake *Oxybelis acuminatus* whose length is approximately 118 centimetres, about a third of which is accounted for by the tail. The maximum breadth of the body is about seven millimetres and the head is very narrow and pointed. When this snake is hanging in the branches, it stretches almost thirty centimetres of the front part of its body out horizontally and waves it to and fro.

A completely separate type of imitation is mimicry. In this case an animal of one species imitates in its coloration an animal of a different, often quite unrelated species. There are two kinds of mimicry. Batesonian mimicry essentially serves to deceive potential enemies by a relatively rare, edible and on the whole defenceless species of animal imitating in its coloration another more common species which is more or less inedible or has some other effective means of protection. On the other hand, Müllerian mimicry involves a number of genuinely harmful species all sharing a similar pattern of warning coloration. This makes it easier for enemies to recognize them as being dangerous or distasteful.

In order that mimicry be effective, however, several conditions must be fulfilled. The first of them is a suitable numerical proportion between the 'imitator' and the 'original'.

With the help of its 'artificial' eye the foureye butterflyfish *(Chaetodon capistratus)* can easily confuse its natural enemies.

119

If an animal is to reinforce, or at least maintain in the enemy the feeling that a certain type of coloration is linked with danger or an unpleasant experience, there must be many more animals that really do possess certain means of protection, than there are of the harmless imitators. A close topographical relationship between the two is just as important. They must not merely have the same geographical distribution, but also the same habitat.

There are not many examples of mimicry among the vertebrates, but one such instance is the imitation of the poisonous weever fish *(Trachinus vipera)* by the harmless saltwater sole *(Solea vulgaris)*. The weever fish has a black dorsal fin and the sole has a similar black patch on the right thoracic fin. This similarity in coloration could be coincidental, but several circumstances bear out the fact of mimicry. Both species have practically the same geographical distribution, they are similar in their way of life and they live against the same coloured background. When excited, *Trachinus vipera* raises its black dorsal fin and when *Solea vulgaris* is alarmed its black thoracic fin stands up erect and opens out so that it resembles the fin of the weever fish. The thoracic fins of other flatfish are not black and they do stand erect in similar situations.

Though this book is about animal coloration, there will be a minor exception in this chapter and mention will be made of the coloration of an animal product — birds' eggs, and in particular the eggs of cuckoos which in their coloration provide an ideal example of the extent to which mimicry can be applied in the world of animals. There are a great number of species of cuckoo in the world, but only some of them place their eggs in the nests of host birds. The others raise their chicks in the normal way. The eggs of these non-parasitic cuckoos are generally white, but in the parasitic species this white colour is replaced by many other colours and hues, and also by various types of pattern which result in a very close imitation of the coloration of the host bird's eggs. Whenever a large number of hosts with eggs of different colours nest close to one another in an area, the colour adaptation of the cuckoo's eggs is not particularly good. However, where the

The bittern *(Botaurus stellaris)* on its nest. The spots along the lower part of the neck look like shadows thrown by reeds and facilitate the bird's concealment.

Animals with camouflage coloration characteristically remain motionless and shut their eyes. The nightjar *(Caprimulgus europaeus)*, whose coloration is similar to that of the bark of trees, also has a distinctive way of perching on a branch. It does not sit across it like most birds, but along the branch.

A close-up of the nightjar see on the opposite page.

cuckoos make use of the nests of only one host bird, the colour adaptation of their eggs is remarkable and may be very far removed from the original type. Some cuckoos specialize in the choice of host to such a degree that for a hundred eggs placed in the nests of regular hosts there is only one cuckoo's egg laid in the nest of a different host.

The European cuckoo *(Cuculus canorus)*, however, is not the only species which leaves the rearing of its young to foster-parents. There are a number of other species of cuckoo which are parasitic and which have also developed mimicry in the coloration of their eggs to a high degree. For instance, the koel *(Eudynamis scolopaceus)* and the great spotted cuckoo *(Clamator glandarius)*, whose young are reared in the nests of birds of the crow family, have blue eggs covered with numerous brown spots. One of the types of egg laid by the cuckoo *Cuculus poliocephalus* is coloured almost red. The eggs of the Asiatic form of the European cuckoo *Cuculus canorus telephonus*, which is parasitic on the bunting in Japan, are covered, especially at the blunt end, with intertwining lines, stripes and spots.

These species of cuckoo, which have a very wide distribu-

tion, lay eggs of two or more colour types in different places. For example, the large hawk-cuckoo *(Hierococcyx sparveroides)*, which has two host birds, lays both olive-coloured and blue eggs. Although this cuckoo and another species with the same ability and roughly the same numbers live in close vicinity to one another in Assam, eggs of a transitional colour have not yet been found, and eggs of the wrong colour have not been found in a nest.

In some species of cuckoo colour adaptation of hatched chicks also occurs. Chicks of the European cuckoo keep the coloration typical of their species and differ in colour from their step brothers and sisters, but in the case of the great spotted cuckoo, whose young are reared in the nests of magpies *(Pica pica melanotus)* and hooded crows *(Corvus corone cornix)*, the young, unlike their parents, have dark heads similar to those of the host bird's young. Similarly, the heads of chicks of the pied crested cuckoo *(Clamator jacobinus)* are lighter in colour than those of the adult birds.

COLOURS AS PART
OF PROTECTIVE EQUIPMENT

It has already been mentioned several times that protective coloration on its own, like animal coloration in general, would not be so effective if it were not accompanied by corresponding behaviour or defence capabilities in its bearer. Many examples of such behaviour are known today.

The chapter dealing with the principles of protective coloration described how in many species the pattern creates a false impression of the position of the body. Some coral fishes such as the four-eye butterflyfishes *(Chaetodon capistratus)* have their eyes camouflaged with a dark stripe, while in the vicinity of the tail each has a large striking 'eye' pattern. The impression of an actual eye is heightened by the fact that the fishes swim slowly with their tails to the front and only when they are disturbed do they make off in the right direction.

Very many interesting observations have been made concerning the behaviour of animals which have camouflage coloration. In the literature dealing with the subject are descriptions of the behaviour of three differently coloured species of iguana of the genus *Anolis* (one brown, one green and one light grey with darkish patches), which occur in the same area, hunt the same insects and on the whole live in the same way. When frightened, however, the animals of each species hide in surroundings that correspond to their own coloration, so that they are completely lost to view.

Motionlessness accompanied by the taking-up of a special

Sex differences in the coloration of the Magellan goose *(Chloephaga picta)*. The male has a black pattern on a pure white background while the basic coloration of the female is brown.

The coloration of the East Asian pigeon *Phlogoenas luzonica* is one of the many examples of coloration which has no explanation. It is impossible to rid oneself of the impression that it is a wounded bird but no logical explanation of the purpose of this coloration has been found.

The coloration of the royal python *(Python regius)* is an example of cryptic coloration based on the dismembering effect of the pattern which renders it difficult to make out the actual shape of the body.

body position is often characteristic of animals with camouflage coloration. Nightjars (Caprimulgidae) sit lengthwise on a branch or press themselves to the ground. The similarly coloured long-eared owl *(Asio otus)* on the other hand stands erect, pricks its 'ears', half closes its eyes and draws in its feathers. Many agamas are not hard to catch, but are very difficult to find. 'If one succeeds in seeing the *Ophyoessa* iguana on a dead branch, it is very easy to catch it in one's hand as it makes no attempt to escape', writes R.W. Hingston, the author of a book about animals in British Guinea. According to other observations the South American variegated heron *(Ardetta involucris)* sat motionless on the stem of a reed, with its beak, head, neck and body all stretched like a continuation of the reed. When the observer walked round it, the bird kept turning the front of its body towards him and allowed him to bend its head right down to its back. When freed the bird returned like a spring to its erect position. Incidentally, the European bittern *(Botaurus stellaris)* has a similar behaviour pattern.

Instinctive motionlessness is especially characteristic of young birds. J.B. Cott describes his encounter with some young woodcocks as follows:

'I was walking one July afternoon with Mr. Richard Elmhirst through a strip of woodland near Millport, Cumbrae, intent on finding nesting woodcock — which I was anxious to photograph. Suddenly, as we neared the edge of the spinney, a woodcock flew up a yard or two in front of my companion. We both went forward to the place, and stooping down, Mr. Elmhirst pointed out to me a small drop of fresh excrement which, left by the rising bird, revealed the exact spot she had just quitted. I mention this trivial circumstance because it was only a moment later that we realized that there, right under our very eyes, were crouching four recently hatched chicks. A spot of watery fluid on a leaf had been easier to recognize than the young birds, so effectively were their fluffy forms broken up by the disruptive pattern of their coats of down.'

Animals with warning coloration also have much in their behaviour that is noteworthy. For example, the Cape polecat

(Ictonyx striatus), an African beast of prey of the stoat family, always shows its enemy the white stripes on its back. J.B. Cott quotes in his book on the protective coloration of animals the observations of several people who came across it on a walk:

'When first seen, the beast appeared to be shamming dead; it lay on its side, with its legs stretched out on the ground towards them. However, on their looking back, having momentarily turned aside, it was noticed that the animal had turned over, so that its back was now directed towards them. Interested in this, they walked round the animal, pretending to take no notice of it, and on again looking round at it, the polecat was found to have turned over, so that once more the stripy back was directed towards them. They repeated the process two or three times and on each occasion the creature turned over, presenting to full view the conspicuous black and white warning livery'.

As already mentioned, these animals actually call attention to their dangerous weapons, or are coloured in such a way that an enemy which has already come into contact with their defense mechanisms will have no difficulty in remembering them. An example of this first type of behaviour is found in the weever fish *(Trachinus vipera)*, which has poisonous spikes on the gill lids in the vicinity of the dorsal fin. Another instance is the similarity in the coloration and behaviour of those species of stoat-like beasts of prey which have huge stink-glands. Their coloration is basically very dark or black, supplemented by one or several white stripes along the back or the head. While motionlessness and inconspicuous behaviour is typical of animals with camouflage coloration, animals with warning coloration characteristically behave naturally or even noisily so as to draw attention to themselves. T. Belt, a nineteenth-century naturalist wrote the following description of an encounter with a young skunk *(Mephitis mephitis)*: 'The only animal we met with was a black and white skunk, with a young one following it. The mother ran too fast up a rocky slope for the young one, which was left behind, and came towards us. It was very pretty, with its snow-white bushy tail laid over its black back. We were, however, afraid to touch it, fearing that, young as it was, it might have a supply of that foetid fluid that its kind discharge with too sure an aim at any assailant'.

Animals which basically have cryptic coloration, but which are capable of colour change use the element of surprise to fend off an attack. The green-coloured East African chameleon *Chamaeleo dilepis* when alarmed, immediately turns black, doubles its size by blowing up its body, shows the bright colour of the inside of its lips and hisses like a snake.

Finally, some animals increase the effectiveness of their coloration by sheer numbers. Here is T. Mortensen's description of his encounter with the small marine catfish *Plotosus anguillaris:* 'A striking black object moving in the shallow water on the coral reef near the island attracted my attention. When I went closer I discovered that it was a mass of little black fishes with two white bars along their backs. The continual movement of the fishes made the whole formation very striking. I wanted to catch several of them and so I began to take some out in my hand. The first fish . . . bit my finger, causing a sharp pain. When I tried to pull if off, it clung to another finger. The pain lasted a long time after I managed to pull it off . . .'

Cephalopods have highly developed eyes which bear a remarkable resemblance to those of vertebrates. The photograph shows squids *(Loligo vulgaris).*

HOW ANIMALS SEE

So far the external manifestations and laws of animal coloration have been described, but it must not be forgotten that coloration has another aspect concerned directly with the lives of animals both in their relationships to their environment and also in their mutual relationships within the community. The problem which hangs like a huge question mark over all that has so far been said about animal coloration, still remains. Does all this really have any significance for the animal?

The basic prerequisite for answering this question is a knowledge of the quality and the limits of the capabilities of animals' visual organs. The manner in which human beings see is not necessarily the same as the way animals see colours and forms and thus in reality a particular coloration may appear quite different to other species from what we imagine. The ability to see, or more exactly the ability to perceive light rays, is present in almost all animals. The degree of perfection of perception and the complexity of the light-receiving organs, however, vary greatly in individual forms.

In most protozoans there is no special equipment for receiving light rays and the cell plasma in general would seem to be sensitive to light. Some of them have a simple colour spot — the stigma. In more complex multicellular organisms several cells dispersed through the skin specialize in this task but these creatures' ability to see is often limited to a differentiation between light and dark as in sponges, coelenterates and also in some worms, such as the earthworm *(Lumbricus terrestris)*.

 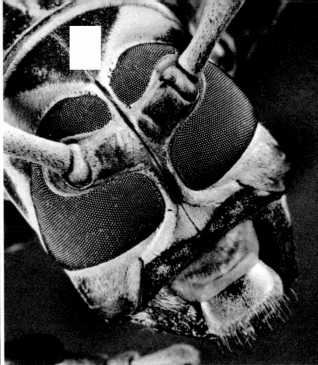

More perfect, directional vision is made possible by the light-sensitive cells being shaded on one side with pigment so that they can only receive light rays from a certain direction. In addition to this they are usually concentrated into largish patches, forming a sort of 'flat' eye. Many invertebrates have these as also does the relatively highly organized lancelet *(Branchiostoma lanceolatum)*, which is a primitive chordate.

The most perfect perception of images is made possible by a special construction of the eyes, which in this case have the appearance of enclosed sacs connected to the external environment by a narrow opening. The most highly organized creatures, principally vertebrates, are capable of this kind of vision, but it also occurs in some specialized groups of invertebrates such as insects (Insecta) and squids and octopuses (Cephalopoda) and it extends relatively far down the scale to primitive animals like some jelly-fish (Scyphozoa) and worms (Annelida).

There are two types of eye capable of distinguishing forms and the delicate structures of a pattern: compound eyes which

On the left is the slender loris *(Loris tardigradus)* showing typical vertebrate eyes. On the right the musk beetle *(Curocinus bunginaeus)* shows the compound eyes of insects.

are typical of insects, and camera eyes which are characteristic of vertebrates.

The compound eye consists of a large number of small cones, each of which is a miniature eye with its own lens, retina and nerve. The total image is seen as a mosaic created by a large number of separate images of parts of the object observed, each of which is registered separately by one of the individual eyes. Compound eyes can only see sharply at close quarters and at a greater distance they merely register movement. They are not capable of accommodation, that is focussing to the varying distances of the objects observed. Therefore in some quickly moving arthropods the eyes consist of two parts: the upper, which is adapted to the observation of movement and contains a smallish number of large eyes, and the lower part, intended for sharp vision at close quarters, which contains a large number of small eyes.

The eyes of vertebrates are typical camera eyes. Among the invertebrates only cephalopods have eyes of similar construction, though they originate in a somewhat different manner. Camera eyes generally have the appearance of an enclosed sphere which is optically connected to the external environment by a small opening — the pupil. Light rays from the object are focussed by the lens and the aqueous humour onto the layer of light-sensitive cells called the retina situated on the internal circumference of the eye. Unlike the compound eye, the camera eye is generally capable of accommodation. Every eye has it own particular field of vision, namely that section of its surroundings which it is able to register at once. This angle is often very large, for example, the angle of vision of the pigeon is 150°, so if an animal's eyes are positioned at the side of the head it can see practically everything that is going on around it.

An exact estimate of the distance of objects is especially necessary for active tree-living creatures such as monkeys, and for owls and carnivorous mammals which hunt live prey. Such species have the ability to see in depth due to the fact that their eyes face forward, so that their fields of vision overlap and an object is observed simultaneously from two slightly different angles.

However, just as important as the actual perception of objects is the consideration of the extent to which animals are capable of perceiving colours and surface structures of objects. Determining these abilities is a much more complex matter. The actual anatomical prerequisites for colour vision are known only in vertebrates and are the presence of two types of cell in the retina of the eye, so-called cones which make possible colour vision, and rods which are necessary for black and white vision.

According to the most recent assumptions, birds, some groups of mammals, tortoises, fishes and apparently even some amphibians are capable of colour vision. Cartilaginous fishes (Chondrichthyes) do not distinguish between yellow and red while the remaining vertebrates are apparently colour-blind and see in black and white. Our knowledge of invertebrates is limited in this field, as even in experiments with different colours one cannot always say with certainty whether the animal is really reacting to the colour or merely to various shades of grey. So far colour vision has been demonstrated experimentally in butterflies and bees. Apart from butterflies, invertebrates do not usually perceive red colours.

In general, one can say that vertebrates see longer light waves better (humans 750—400 mμ), while colour vision in invertebrates is shifted more to the ultra-violet range of the light

133

spectrum (honey bee 600—300 mμ, water fleas 600—200 mμ).

The ability to see either colours or black and white has yet another aspect. Animals which can see colours can see very well when it is light, but in the evening and at night they are almost helpless. On the other hand, animals that see only in black and white can distinguish even small differences in light intensity and can move around when the absolute light intensity is minimal. It is reported that the rods which are responsible for black and white vision are a hundred times more sensitive than cones.

Of course, one must remember that even animals with colour vision usually have a number of rods in their retinas which enable them to see to a certain extent in twilight.

There is another specialization that occurs in animals which live permanently or mainly in an environment with little light. This is an overall enlargement of the eye. In the strictly nocturnal tawny owl *(Strix aluco)* the weight of the eyes amounts to a thirtieth of the total weight of the body, while in the equally large diurnal relatives such as the short-eared owl *(Asio flammeus)* they amount to only one ninetieth of the weight of the body. In some cephalopods which live at great depths the diameter of the eyes is as much as forty centimetres!

Yet another adaptation which makes use of weak light intensity is the creation of a special shiny layer positioned under the retina, which reflects the light rays that fall on it and thus enhances their effect. This layer generally consists of guanine crystals or golden granules of riboflavine. It is found in the eyes of members of the shark family (Elasmobranchii), the sturgeon *(Acipenser sturio)*, some teleost fishes (Teleostei), reptiles, birds and many mammals, mainly beasts of prey and ungulates, and it causes the luminescence in these animals' eyes so beloved of makers of cartoon films. This luminescence does not originate in the eyes but is merely a reflection of the rays entering them.

The question of vision is, of course, more complex than might be assumed after reading this chapter, both as regards the construction of the visual organs and the theory of vision. Little has been said about sharpness of vision, which varies greatly in different species and groups. The visual abilities of birds of prey are as many as five times better than those of humans, whereas many other animals cannot, from a greater distance, distinguish objects which are not moving. Not even the construction of the visual organs has been dealt with in detail: for instance, some so-called four-eyed fishes of the genus *Anableps* have eyes divided into two halves, one of which is adapted to seeing in the air and the other to vision in water.

Not even opinion as to the question of colour vision would seem to be completely undivided. In general zoologists accept that most mammals are not capable of perceiving colours and this is associated with the lack of bright colours in their coloration. However, some physiologists believe that colour-blindness affects only a small section of mammals. There are many such differences of opinion, but they extend beyond the framework of this chapter, the sole aim of which is to provide the reader with an idea of the visual abilities of animals.

The polar bear *(Thalarctos maritimus)* has no enemies in its habitat. The white coloration makes it easier for him to approach his prey unnoticed.

THE EFFECTIVENESS
OF COLORATION

Several chapters have already been devoted to a theoretical explanation of the principles of protective coloration. The question remains, however, as to whether this theoretical explanation is correct and whether these principles are actually applicable in the world of animals.

One is justified in suggesting that having the same colour as the ground, or in fact, having any camouflage coloration cannot be of much use to animals whose main enemies are beasts of prey or owls. These predators hunt primarily by smell and hearing, senses against which coloration would provide no protection at all. In the same way it can be argued that there are many species which might be expected to be camouflaged and yet which have conspicuous coats. In arctic regions there live dark-coloured wolverines *(Gulo gulo)* and dark musk-oxen *(Ovibos moschatus)* in addition to white polar bears *(Thalarctos maritimus)*, arctic hares *(Lepus timidus)* and arctic foxes *(Alopex lagopus)*.

If logical considerations are rejected as being insufficient proof of the correctness of the theory, one must in practice fall back on the results of various observations and experiments in order to challenge such objections. These observations include the reaction of carnivores to prey of different colours and the relative amount of this prey in their catches.

The diet of predators which hunt by sight has been investigated in experiments using the mosquitofish *(Gambusia*

The white-tailed porcupine (*Hystrix leucura*). The striking black and white coloration supplements the intimidating effect of the spines which are raised when the porcupine is disturbed.

patruelis), which is capable of changing its colour to a considerable degree. For a period of about eight weeks the fishes were kept in tanks with either black or white walls. In the first instance the back and flanks of the fishes turned almost black and the second group became light cinnamon. Then both groups were transferred to black and light grey tanks and various fish-eating birds and preying fishes were given access to them. Of the total number of 542 fishes eaten, only one-third were individuals already adapted in colour to their environment, whilst the remaining two-thirds consisted of fishes differing in coloration from their environment.

Other observations concern the prey of three ravens which for a period of two hours hunted over two and a half acres of pasture land. Here 300 young chickens were gathering food; 120 of them were black, 120 white and the remainder were coloured individuals similar in their coloration to the red jungle fowl *(Gallus gallus).* Only one coloured chicken occurred in the ravens' catch, the remainder consisting of ten white and thirteen black chickens.

This observation recalls another interesting field observation. J.S. Huxley cited evidence that in Scandinavian willow ptarmigan *(Lagopus lagopus)* the moulting of white winter feathers by the male in spring is postponed until the female, in her brown summer camouflage plumage, has finished hatching her eggs. He considered this to be a protective adaptation drawing attention away from the more important female during the nesting period to the less important male, which as a result of his more conspicuous appearance more frequently became the quarry of beasts of prey.

Even more interesting results are provided by experiments concerning warning coloration, which would seem to be much more effective than camouflage coloration. For instance, of thirty insects with camouflage coloration which were presented as food to birds, only four were uneaten (13.3 %), while of twenty-eight with warning coloration twenty-five (89.3 %) escaped serious attack.

Prey with camouflage coloration accounted for 86 % of the consumption. of nesting birds compared with 9 % of species with warning coloration. Analyses of the natural nourishment of frogs in east Africa revealed that from a total of 11585 examples of prey only 0.17 % had typical warning coloration.

Dietary analyses are a relatively reliable means of determining the significance of several aspects of animal coloration. Certain results, however, are acquired by other methods such as direct observation. Marion Hubbard gave a Californian newt *Diemyctylus torosus*, which is coloured like the European salamander, to the snake *Thammophis elegans* to eat. The newt was presented every day but only after eleven days without any nourishment did the snake attack and then it almost immediately let go of the newt again, opening its jaws as if it wanted to get rid of the taste of the prey. Its experience had such a profound effect that in three further experiments the snake did not touch the newt at all. It is interesting that the European salamander would seem to have similar protection, as it has not been discovered in the diet of birds. Even though the primary reason is no doubt the poisonous secretion of the skin glands, this experiment bears witness to the great effectiveness of warning coloration.

There are also observations which would seem to indicate that coloration has a certain protective value, even against animals in which senses other than vision are important. This is significant, especially as those who doubt the effectiveness of protective coloration very often argue that animals hunting prey find their orientation primarily through senses other than sight. Even though these senses (smell, hearing) are very acute, it would seem that visual clues are absolutely necessary for the execution of the final attack. This is shown in observations made of the ribbon snake *Eutenia saurita* catching a frog. The snake pursued it quickly, but when the frog stopped on ground covered with leaves the snake halted about two and a half metres from it and observed the surroundings with its head erect and its tongue out. Only when the frog moved again was it seen and caught.

The above-mentioned experiments and observations are of course not sufficient to confirm or challenge the effectiveness of protective coloration. Apart from this, one must also take into account the fact that exceptions can always be found to an accepted rule. It is just the same with the question of protective coloration. Almost always there occur specialized enemies against which protective coloration is of no value, just as there are others which are not bothered by the hard bristles of caterpillars, the spines of porcupines or snakes' poison. But there is really no doubt that protective coloration is one of an animal's means of defence. In the view of J.B. Cott, who studied these problems in great detail, the function of protective coloration is not the absolute protection of its bearer, rather it acts together with the animal's other means of defence and increases their effectiveness. It does not occur generally and it may be completely lacking in animals which do not require it because they have other means of defence. This fact does not disprove the protective function of coloration, nor do the dark colours of some polar animals and other apparent illogicalities which anyone who examines these questions in depth will encounter.

COLORATION FOR COMMUNICATION

If communication between animals is considered to be an automatic reflex release of certain signals which result from the overall state of the animal, one should include among the means of communication some types of coloration and some reactions, whether they be the sudden change in colour of chameleons, or the sudden exposure of a strikingly coloured

part of the body or any other form of demonstration. The same is true of colour changes which express a certain state in the animal. For instance, among fishes anger or preparations to attack are often associated with a darkening of their colour and when frightened they turn a lighter colour.

Some changes in animal coloration, especially in males during the reproduction period, also appear to have a signalling function. Mention has already been made of the changes in coloration of squirrels and many fishes, which are quite obviously connected with the courtship behaviour of males.

But when coloration is considered as a means of communication between animals, one thinks primarily of colour characteristics which serve exclusively as a specific signal, either permanently, or linked with a certain type of behaviour which makes the coloured areas especially conspicuous. Such areas may be positioned on various parts of the body such as on the wings of ducks and pigeons, on the fins of fishes and in the region of the anus of ruminants. The position depends to a considerable extent on the role they play in the life of the animal.

Some of these signal patterns direct attention to the animal or enable it to be recognized at a great distance. They consist of a simple pattern of clear colours which covers a largish or con-

The signal spots of mammals may be positioned on various parts of the body: on the rear part of the body in the region of the anus [left — the springbok *(Antidorcas marsupialis)*], on the limbs [centre — the giant eland *(Taurotragus derbianus)*], or on the head and neck [right — the nilgai antelope *(Boselaphus tragocamelus)*].

141

spicuous part of the body. These are, for instance, various white patches exposed by birds when in flight like the white rump of the bullfinch *(Pyrrhula pyrrhula)*, the white tail feathers of many songbirds, the speckled wingtips of the hoopoe *(Upupa epops)*, the white patches of roedeer and many other even-toed ungulates, and the brick red breast of the robin *(Erithacus rubecula)*. In mammals this coloration generally appears at both extremes of the body; on the head, in the form of various masks which either facilitate identification or show readiness to fight, and at the anal end of the body in the form of 'mirrors' which have a sexual function or act as a warning signal.

A second group consists of signals which give rise to emotions, including fright, sexual excitement and so on. These signals are generally smaller, more colourful and have a more complex pattern, and are used by animals at close quarters. They include various coloured 'mirrors' on the wings of ducks which are mainly displayed during mating, the coloration of the neck collar of amherst pheasants *(Chrysolophus amherstiae)*, the pattern on the neck of the cobra *(Naja naja)*, and the coloration of the inner side of the thighs of the diana monkey *(Cercopithecus diana)*.

The coloration of the ears of many mammals is also interesting. For instance, on the rear of the ear lobes of most preying cats there is a clear pattern, largely unrelated to the coloration of the rest of the body. In the species that have patterns on their bodies, the colouring of the ears is black with a striking white patch, as in the tiger and the salt cat *(Oncifelis geoffroyi)*. In single-coloured species like the puma *(Puma concolor)*, the lion *(Pathera leo)*, and the caracal *(Caracal caracal)*, the white colour is generally lost, but the black characteristics are preserved to a greater or lesser extent. In the caracal the whole rear side of the ears is a deep black colour. It is most probable that this pattern on the ears has a signal function.

Differences in the coloration of the minnow *(Phoxinus phoxinus)* at rest (left) and when excited (right). (After Meyer-Holzapfel.)

In this chameleon-like Cuban reptile *Chamaeleolis chamaeleontides* the eyelids continue the pattern of neighbouring regions of the head and help to conceal the eye. The light patches on the body imitate the shape and colour of lichens such as are found on the trunks of trees.

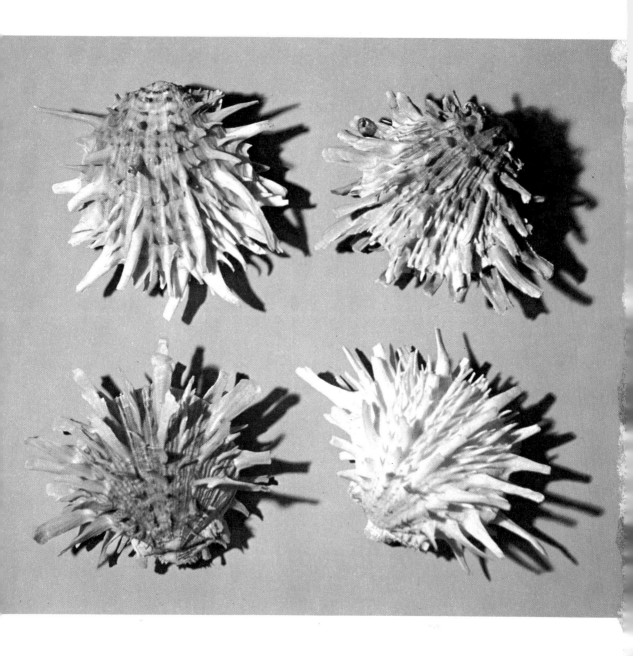

An example of the variability of the Californian mollusc *Spondylus pictorum*. In nature the coloration of this species varies from pure white through various shades of pink, ochre and red to dark violet.

especially as movement of the ears is for many mammals a very noticeable part of their overall behaviour.

Some colour signals are only displayed for a short time and these are indicative of the momentary physiological or psychological state of the animal. Such sudden changes are possible in the lower vertebrates and only exceptionally in birds and mammals. For example, the change in colour of the wattle of the turkey *(Meleagris gallopavo)* or the reddening of the ears of the Tasmanian devil *(Sarcophilus harrisii)* can take place rapidly. Other vertebrates achieve a similar effect by exposing patches of colour which normally remain concealed. For instance, the barasingha *(Cervus duvauceli)* has a relatively long tail, coloured dark on top, which hangs down freely when

The Malayan tapir *(Tapirus indicus)* also has a conspicuous white pattern on its ears (bottom left).

The male of the diana monkey *(Cercopithecus diana)* can startle his enemy by suddenly opening his thighs and showing the light colour of the inner sides.

at rest. When running away, however, the animal raises it vertically in the air, so that the bright white coloration of the underside is displayed and acts as a signal.

It is an interesting fact that a specific coloured patch is sometimes more important to animals than their overall appearance. When experiments were carried out to find the way in which birds defend their territory, it was discovered that red is such a signal to the male robin *(Erithacus rubecula)*. The owner of the territory not only attacked another robin which appeared there, but was just as aggressive towards a mere tuft of red-coloured cotton wool. And yet he did not react to young birds which had not yet acquired red breasts.

Red appears to be exceptionally important for many birds. This is borne out both by the fact that various shades of red very frequently appear in their coloration and also by the colour of the flowers which the birds visit. For instance, of 110 species of flowers observed to be visited by hummingbirds, forty-five were coloured red, nineteen purple and fifteen orange. This proportion is apparently related to the construction of birds' visual organs and their ability to perceive red light rays.

On the other hand, for most other animals the contrast between colours, often between black and white, is a much more important visual signal. This applies primarily to mammals which in general lack the ability to see colours and for whom differences in various shades of grey are of greater importance. For species with daytime activity black would seem to have the greatest signalling significance. White, which is appropriate to nocturnal species often appears in their coloration, as for example in many beasts of prey of the stoat family.

Red as an optical signal is displayed in a most interesting way in the relationship between parents and young. Many young birds have bright red or orange-coloured insides to their beaks often with yellowish orange 'lips' at the edges. Though one cannot say to what extent this coloration is of directional significance and to what extent there are other reasons for it being there, it has without any doubt a signalling function.

The function of red as an important signal characteristic has been displayed in experiments to determine how young

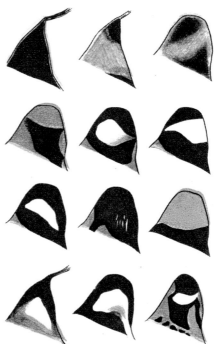

The coloration of animals' ear lobes has a function as a means of communication and may also be the most conspicuous aspect of the animal as in the case of the Siberian tiger, *(Panthera tigris altaica)*, a photograph of which is shown here on the left. The diagram shows the variations in pattern of the right ear lobes of preying cats: 1 the caracal, 2 the swamp cat, 3 the puma, 4 the lion, 5 the jaguar, 6 the leopard, 7 the tiger, 8 the clouded leopard, 9 the cheetah, 10 the lynx, 11 the salt cat and 12 the serval.

birds recognize that their parents, and not some alien creatures, are approaching the nest. N. Tinbergen, the well-known ethologist, carried out these experiments on young herring gulls and discovered that the orange-red spot on the lower jaw of the parent's broad yellow beak is such a signal for recognition. He made models of an adult gull's head, on which the beak and the spot were painted in different colour combinations. When these were held out the young birds reacted to red most of all. It was even demonstrated that it was not the shape of the parent's head, but the colour of the beak which was of most significance, for the chicks reacted most of all to an ordinary stick with red stripes painted on it.

This is not the only sort of identifying signal between young and parents. In the chapter on age differences mention was made of an interesting white pattern in the coloration of some young mammals and birds which distinguishes them from their otherwise similarly coloured parents. It is possible that these are signals, though more exact and systematic observations need to be made before it can be proved.

COLOUR AND PATTERN
IN THE ANIMAL KINGDOM

The number of vertebrates now living on Earth is estimated at nearly fifty thousand species. This book has only referred to a few score. They are just samples selected in such a way as to give the reader at least some idea of the coloration of vertebrates taken as a whole.

The number of vertebrates, however, forms only an insignificant part of the million or more species of which the present-day world of animals consists. What is coloration like in the other groups? It has not been possible to mention them in detail here, but it would be just as impossible not to mention them at all, not only because they form an organic part of the living world around us, but also because the question of coloration cannot be restricted to a certain group. Family relationships do not play an important role here. In mammals and fishes the same principles of pattern can be found, whereas completely different types of coloration exist between closely related species.

Even though analogies might be expected between the coloration of vertebrates and invertebrates it is still surprising to find how frequently they occur and how detailed the correspondence of pattern may be between such heterogenous animals as molluscs and mammals, caterpillars and fishes. It is easy to forget that such a correspondence is only an external similarity which is in some cases neither indicative of any close relationship nor a result of the functional convergence.

This book began by asking how coloration is produced and the same question can be asked again now in a wider context. In invertebrates coloration is caused by pigments, many of which are similar to those found in vertebrates, although some, such as the white leucopterin and the yellow xanthopterin of Lepidoptera, are distinctive of a particular group. As in vertebrates, a metallic lustre is produced on the surface structures through interference between light rays, and a similar coloration of the chitinous structures occurs in some arthropods. Typical are the wingcases of beetles, the wings of some butterflies and moths, and the bodies of ruby-tail wasps (Chrysididae) and some flies. The lower vertebrates have iridocytes which cause a silvery lustre and so do some molluscs, in particular cephalopods. Not only transparent fishes are known, but also transparent invertebrates such as some crustaceans and jellyfishes (Scyphomedusae).

Similar patterns of coloration in the wing feathers of the argus pheasant (*Argusianus argus*), left, and the underside of the wings of the South American butterfly *Calligo eurylochus*, right. It is interesting that in both cases the outer part of the 'eye' is a white pattern which helps to create the illusion of light being reflected from a spherical body.

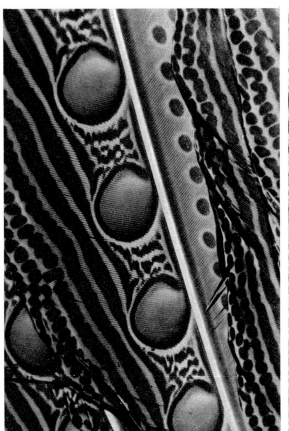
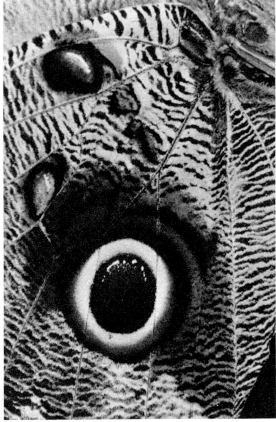

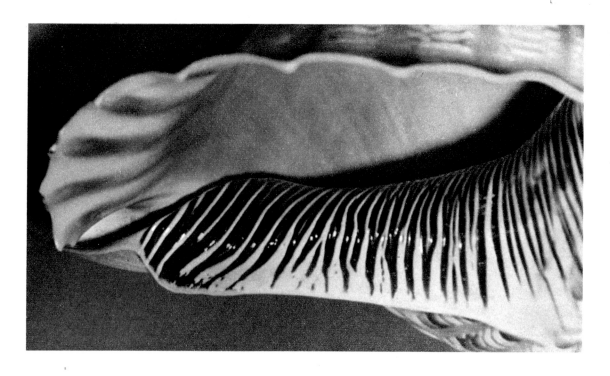

The pattern of invertebrates is surprisingly bright and
varied, even in such animals as tropical flatworms (Turbellaria)
and various molluscs where one would hardly expect it. Some
photographs have been selected to show interesting and striking
parallels with the patterns of vertebrates.

The variability in the coloration of some groups of
invertebrates is enormous and exceeds that known in vertebrates. The variability is both geographical and individual.
For instance, the coloration of extreme forms of the wood tiger
moth *(Parasemia plantaginis)* varies so much that it gives the
impression of negative and positive versions of one and the
same pattern. While some animals have a black pattern on
a light yellow ground, the others have a yellowish pattern on a
brownish black ground. The colour of the front wings of the
black arches moth *(Lymantria monacha)* also varies from a light
shade with darkish spots to an almost monochrome brown.

Like the lower vertebrates, some invertebrates are also
capable of quick colour change. Among cephalopods octopuses
are well-known for this ability. The speed and gradation of

colours in their colour changes are amazing and occur both when the animals are aroused and when the colour of the background against which they are moving changes. It has been observed, for example, that dark-coloured cuttlefishes can contract the pigment in their chromatophores in a fraction of a second so that they become quite pale.

Coloration anomalies are less striking in invertebrates than in vertebrates, no doubt because the animals themselves are not so conspicuous, but some variations such as different temperature forms of Lepidoptera (produced by either heating the pupae to 40° C—43° C or cooling them to below freezing point) are well known. Also worthy of note is the so-called 'industrial melanism' of moths. This is found in heavily industrialized districts and originates as a permanent adaptation to the dark smoky environment. It affects particularly the lighter-coloured species which become very conspicuous in an environment polluted by industrial grime.

Thus for practically each chapter one can find parallels between the coloration of vertebrates and invertebrates, regard-

The distinctive pattern of the alpine musk beetle *(Rosalia alpina)* consists of black spots on a greyish blue ground.

less of whether one is dealing with heredity in coloration — a standard subject for genetic experiments is the vinegar-fly *(Drosophila melanogaster)* — or age, sex or seasonal differences. The parallels in the age differences in the coloration of invertebrates, for instance, are the changes in coloration from the larval stages to adulthood in many groups of invertebrates. Striking sexual differences in coloration are found in many dragonflies (Libellulidae) and damselflies (Calopterygidae), and also in beetles, butterflies such as the whites (Pieridae) and tussock moths (Lymantriidae). The map butterfly *(Araschnia levana)* is well-known for an alternation in generations which is somewhat comparable to the seasonal changes in coloration of vertebrates. The spring generation is reddish brown with black spots, while the summer generation is dark with a white pattern.

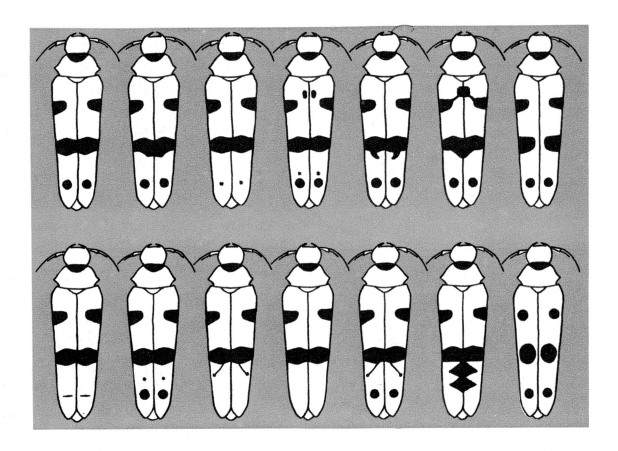

A great number of parallels are to be found in protective coloration. The combination of yellow or orange and black often occurs both in the warning coloration of vertebrates and in many species of invertebrates. It is seen in the common wasp *(Vespula vulgaris)*, the hornet *(Vespa crabro)*, the yellow-fronted digger wasp *(Scolia flavifrons)*, the caterpillar of the cinnabar moth *(Callimorpha jacobeae)* the chrysalid of the magpie moth *(Abraxas grossulariata)*, in many hoverflies (Syrphidae), the cicada *Heteronotus armatus* and many other species.

Similar parallels occur in other types of coloration and in their function. The element of surprise is exploited, for instance, by some locusts and moths such as the red underwing *(Catocala nupta)*. Below the inconspicuously coloured front wings are another pair of wings with a vivid red, yellow or blue colouring.

The shape and position of the spots of the alpine musk beetle are not constant and are even subject to great variability in beetles coming from the same locality. (Diagrams after Korbel.)

The bright pattern of one claw of the fiddler crab functions as a colour signal similar to that described in the chapter on intercommunication between vertebrates.

Apart from these analogies there are of course many types of invertebrate coloration which have not been encountered among the vertebrates, or which are less common. An example is the spider, which in its coloration and behaviour imitates fresh bird droppings sticking to a leaf. A large number of butterflies, moths, locusts, bugs and many other invertebrates imitate leaves or offshoots of plants. The salt-water cuttlefish camouflages itself by emitting into the water a dark pigment which remains suspended there acting as both a 'smoke-screen' and a decoy.

The invertebrates also include relatively numerous species which emit visible radiations and this, to a certain extent, may be considered a form of coloration. But all these are things which would require a separate study. The aim of this chapter is merely to show that what has been written about vertebrate coloration can be extended to invertebrate animals and that in spite of all the great variety in the animal world the same types of coloration and the principles of pattern do recur. This proves that animal coloration in not just a chance combination of colours and configurations, but has firm foundations and a fixed order based on a long evolution.

The pattern on the back of the neck of the Indian cobra *(Naja naja)*, left, and the pattern on the thorax of the death's head moth *(Acherontia atropos)*, right, have much in common, both as regards their position and their shape.

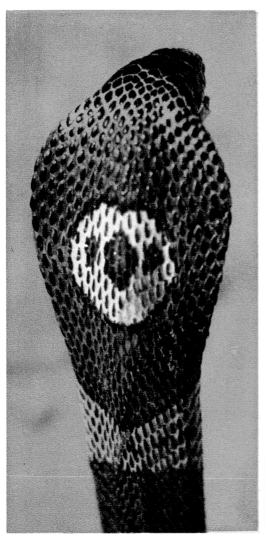

UNANSWERED QUESTIONS

In the preceding chapters an attempt has been made to investigate the origin of the coloured and ornamental beauty of the living world. But is this greater knowledge, and hopefully a deeper awareness of its significance, enough? After all, only a minute section of the wealth of colours in which the natural world abounds has been described and though the subject has been presented logically here, to facilitate comprehension, nevertheless Nature itself seldom offers purely logical phenomena, but continually produces situations which cannot be explained.

Why, for instance, does one find letters resembling those of the Greek and Roman alphabets in the wing patterns of the comma *(Polygonia C — album)* the silver Y moth *(Plusia gamma)* and the tau emperor moth *(Aglia tau)*? What importance to the butterflyfish *(Holacanthus semicirculatus)* is the pattern on its tail in which human imagination has discovered a similarity with ancient Arabic letters which give an Islamic quotation? And why do a number of other animals have in their pattern elements such that people see the shape of a skull on the thorax of the death's head moth, the portrait of a samurai on the crown of the *Dorippe* crab or even the coat of arms of a certain aristocratic family on the heads of Japanese ice fishes *Salanx latus*?

Of course this is just fantasy! But leaving aside the vagaries of human imagination, there are still many questions to be

answered. How it is possible, for instance, that the pattern on various parts of the legs of some frogs is traced in such a way that, when the leg is folded it produces an unbroken line? What produces the shadowing in the pattern of the argus, a flat ornament which creates the illusion of a series of little spheres on which light rays are reflected? What do salt-water fishes of the family Scaridae have in common with parrots whose heads they imitate so fantastically both in their coloration and in the whole shape of the front part of their bodies? Or why does the pattern on the triton's shell correspond so exactly to the stripes of a zebra, when throughout its life it remains covered with a leathery mantle and is only revealed after its death when the soft parts of its body perish?

To find an answer to these questions would mean gathering much more knowledge not only from investigations of the behaviour and distribution of many other animals but also from the fields of anatomy, physiology and genetics, for information from all these fields contributes to an understanding of animal coloration. Perhaps one day the answers will be known. But not even new knowledge will reduce in any way our amazement at the coloured beauty of living creatures and the incredible ingenuity with which it is related to the whole system of characteristics and abilities which enable animals to exist in the multifarious conditions of our Earth.

The material for this book has been compiled from various literary sources — specialist articles, handbooks and also some more complex scientific publications. Of all these, the following few at least should be mentioned:

J. B. Cott — Adaptive Coloration in Animals;
E. Florey — Lehrbuch der Tierphysiologie;
H. Frey — Bunte Welt im Glase;
K. Hrubý — Genetics (in Czech);
I. Krumbiegel — Biologie der Säugetiere;
A. Portman — Einführung in die vergleichende Morphologie der Wirbeltiere;
A. G. Searle — Comparative Genetics of Coat Colour in Mammals;
J.L.R. Smith — The Sea Fishes of Southern Africa;
G. Tembrock — Verhaltensforschung.

The sources from which the models for line drawings and illustrations have been taken are quoted directly in each individual caption.

AUTHORS OF PHOTOGRAPHS

(the numbers refer to pages)

Prof. Dr. I. Eibl-Eibesfeldt (Seewiesen)   108 above
Dr. S. Frank (Prague)   108 below
H. J. Franke (Gera)   114
Dr. H. Hemmer (Mainz)   65
Dr. I. Heráň (Prague)   4, 14, 16, 17, 18 below, 20, 21, 23, 29, 32, 33, 34, 35, 36 below, 37, 40, 41, 44, 47, 49, 50, 57, 63, 64, 68, 69, 70, 73, 78, 80, 83 left, 87, 89 below, 94, 96, 97, 102, 103, 105, 106, 107 above, 110, 111, 125, 133, 135, 137, 140, 141, 144, 145, 147, 149, 150, 151, 154, 157
Prof. Dr. W. Huber (Bern)   89 above
M. Chvojka (Prague)   43, 74, 75, 84, 95 right
Dr. J. Jeník (Prague)   113
Dr. H. Meyer-Brenken (Obernkirchen)   56
A. Olexa (Prague)   42, 126
Ing. W. Puchalski (Kraków)   26, 27
Dr. V. J. Staněk (Prague)   18 above, 19, 36 above, 48, 53, 54, 71, 79, 81, 82, 86, 90, 95 left, 101, 107 below, 109, 112, 118, 121, 122, 123, 129, 131, 143, 152, 155
Dr. Z. Veselovský (Prague)   28, 38, 46, 51
Dr. J. Volf (Prague)   66, 83 right

The photographs on pages 93, 98 and 117 have been adapted after the following authors:
Meerwarth and Soffel — Lebenbilder aus der Tierwelt (1912)   93
Berger — Belauschte Tierwelt (1930)   98
Brehm — Tierleben (1914)   117

Illustrations and drawings Dr. I. Heráň.